Introduction to
Stochastic Dynamic Programming

This is a volume in
PROBABILITY AND MATHEMATICAL STATISTICS

A Series of Monographs and Textbooks

Editors: Z. W. Birnbaum and E. Lukacs

A complete list of titles in this series appears at the end of this volume.

Introduction to Stochastic Dynamic Programming

Sheldon Ross
University of California
Berkeley, California

ACADEMIC PRESS

A Subsidiary of Harcourt Brace Jovanovich, Publishers

New York   London

Paris   San Diego   San Francisco   São Paulo   Sydney   Tokyo   Toronto

ACADEMIC PRESS, INC.
111 Fifth Avenue, New York, New York 10003

United Kingdom Edition published by
ACADEMIC PRESS, INC. (LONDON) LTD.
24/28 Oval Road, London NW1 7DX

Library of Congress Cataloging in Publication Data

Ross, Sheldon M.
    Introduction to stochastic dynamic programming.

    (Probability and mathematical statistics)
    Includes bibliographies and index.
    1. Dynamic programming. 2. Stochastic programming.
I. Title. II. Series.
T57.83.R67 1982   519.7'03   82-18163
ISBN 0-12-598420-0

PRINTED IN THE UNITED STATES OF AMERICA

83 84 85 86   9 8 7 6 5 4 3 2 1

To Celine

Contents

Appendix: Stochastic Order Relations

Preface

This text presents the basic theory and examines the scope of applications of stochastic dynamic programming. Chapter I is a study of a variety of finite-stage models, illustrating the wide range of applications of stochastic dynamic programming. Later chapters study infinite-stage models: discounting future returns in Chapter II, minimizing nonnegative costs in Chapter III, maximizing nonnegative returns in Chapter IV, and maximizing the long-run average return in Chapter V. Each of these chapters first considers whether an optimal policy need exist—presenting counterexamples where appropriate—and then presents methods for obtaining such policies when they do. In addition, general areas of application are presented; for example, optimal stopping problems are considered in Chapter III and a variety of gambling models in Chapter IV. The final two chapters are concerned with more specialized models. Chapter VI presents a variety of stochastic scheduling models, and Chapter VII examines a type of process known as a multiproject bandit.

The mathematical prerequisites for this text are relatively few. No prior knowledge of dynamic programming is assumed and only a moderate familiarity with probability—including the use of conditional expectation—is necessary. I have attempted to present all proofs in as intuitive a manner as possible. An appendix dealing with stochastic order relations, which is needed primarily for the final two chapters, is included. Throughout the text I use the terms *increasing* and *nondecreasing* interchangeably.

1

Finite-Stage Models

1. Introduction

A problem typical of those with which we are concerned involves a process that is observed at the beginning of a discrete time period to be in a particular state. After observation of the state, an action must be chosen; and based only on the state at that time and the action chosen, an expected reward is earned and the probability distribution for the next state is determined. The problem of interest is to choose a policy that maximizes the expected value of the sum of the rewards earned over a given finite time span of length n.

We present a technique, known as dynamic programming, that enables such problems to be solved recursively in n. To be specific, suppose that the states are the integers, and let A, a finite set, be the set of all possible actions. When the state is i and the action $a \in A$ is chosen, suppose that the reward earned is $R(i, a)$ and the next state is j with probability $P_{ij}(a)$. Let $V_n(i)$ denote the maximum expected return for an n-stage problem that starts in state i.

When $n = 1$, we clearly have

$$V_1(i) = \max_{a \in A} R(i, a). \tag{1.1}$$

Also, the one-stage optimal policy is to choose, when in state i, an action that maximizes the right side of Eq. (1.1). Now consider the n-stage problem that starts in state i. If action a is initially chosen, then reward $R(i, a)$ is received and the next state will be j with probability $P_{ij}(a)$. If the next state is j, we then face a problem equivalent to one that starts in j and has $n - 1$ time periods to go. Hence the best we can do (in the sense of expected return) if we initially choose action a is

$$R(i, a) + \sum_j P_{ij}(a)V_{n-1}(j).$$

Because $V_n(i)$ is the best we can do without any restriction on the initial action a, we see that

$$V_n(i) = \max_a \left[R(i, a) + \sum_j P_{ij}(a)V_{n-1}(j) \right]. \tag{1.2}$$

Equation (1.2), known as the *optimality equation*, gives us a technique for recursively solving for $V_n(i)$. First we obtain $V_1(i)$ from Eq. (1.1). Letting $n = 2$ in Eq. (1.2), we can now solve for $V_2(i)$, and so on. In addition, the optimal policy is as follows: when there are n time periods to go and the process is in state i, then an action a that maximizes the right side of Eq. (1.1) should be chosen. (That such a policy is optimal, achieving the expected return $V_n(i)$ when the initial state is i and the problem is over n stages, is easily seen by induction on n.)

It often takes a great deal of computation to solve for the optimal policy explicitly. However, we can occasionally use Eq. (1.2) to solve for V_n explicitly or to obtain structural results about it (or about the optimal policy), which can result in reducing the amount of necessary computations.

In this chapter we shall present and obtain some results about a variety of finite-stage sequential-decision models.

2. A Gambling Model

At each play of the game a gambler can bet any nonnegative amount up to his present fortune and will either win or lose that amount with probabilities p and $q = 1 - p$, respectively. The gambler is allowed to make n bets, and his objective is to maximize the expectation of the logarithm of his final fortune. What strategy achieves this end?

Let $V_n(x)$ denote the maximal expected return if the gambler has a present fortune of x and is allowed n more gambles. We shall take as our action the fraction of the gambler's fortune that he bets. Thus we have the optimality equation

$$V_n(x) = \max_{0 \le \alpha \le 1} [pV_{n-1}(x + \alpha x) + qV_{n-1}(x - \alpha x)], \qquad (2.1)$$

with the boundary condition

$$V_0(x) = \log x.$$

We leave it as an exercise for the reader to show that when $p \le \frac{1}{2}$, $V_n(x) = \log x$ and the optimal strategy is always to bet 0. Suppose $p > \frac{1}{2}$. Using $V_0(x) = \log x$, we have by Eq. (2.1) that

$$V_1(x) = \max_\alpha [p \log(x + \alpha x) + q \log(x - \alpha x)]$$

$$= \max_\alpha [p \log(1 + \alpha) + q \log(1 - \alpha)] + \log x. \qquad (2.2)$$

But simple calculus shows that the maximum of Eq. (2.2) is attained at $\alpha = p - q$, and thus

$$V_1(x) = C + \log x,$$

where

$$C = \log 2 + p \log p + q \log q.$$

Using Eq. (2.1) with $n = 2$ now yields

$$V_2(x) = \max_\alpha [p \log(x + \alpha x) + q \log(x - \alpha x)] + C.$$

Hence, upon comparison with Eq. (2.2), we see that the optimal decision is again to bet the fraction $p - q$ of one's fortune, and

$$V_2(x) = 2C + \log x.$$

Indeed, it is easy to see that

$$V_n(x) = nC + \log x,$$

and the optimal action is always to bet the fraction $p - q$ of one's present fortune.

3. A Stock-Option Model

Let S_k denote the price of a given stock on the kth day ($k \geq 1$) and suppose that

$$S_{k+1} = S_k + X_{k+1} = S_0 + \sum_{i=1}^{k+1} X_i,$$

where X_1, X_2, ... are independent and identically distributed with distribution F (having finite mean) and are also independent of S_0, the initial price. This is known as the *random-walk* model for stock prices.

Suppose now that you own an option to buy one share of the stock at a fixed price, say c, and you have N days in which to exercise the option. You need never exercise it, but if you do at a time when the stock's price is s, then your profit is $s - c$. What strategy maximizes your expected profit?

If we let $V_n(s)$ denote the maximal expected profit when the stock's price is s and the option has n additional days to run, then V_n satisfies the optimality equation

$$V_n(s) = \max\left[s - c, \int V_{n-1}(s + x)\, dF(x) \right], \qquad n \geq 1, \qquad (3.1)$$

with the boundary condition

$$V_0(s) = \max(s - c, 0).$$

There is no way to obtain a simple explicit solution for V_n. However, we can obtain a simple property of $V_n(s)$ that will yield the structure of the optimal policy.

Lemma 3.1 $V_n(s) - s$ is decreasing in s.

Proof: The proof is by induction on n. It is immediately obvious that $V_0(s) - s$ is decreasing in s, so we can assume the same for $V_{n-1}(s) - s$. Now, from Eq. (3.1)

$$V_n(s) - s = \max\left\{ -c, \int \left[V_{n-1}(s + x) - (s + x) \right] dF(x) + \mu_F \right\},$$

where μ_F is the mean of F. By the induction hypothesis,

$$V_{n-1}(s + x) - (s + x)$$

is, for each x, decreasing in s, therefore the result follows. $\square$

Remark: It is obvious (and immediate by induction) that $V_n(s)$ is increasing† in s. We shall also require (and leave the proof as an exercise) that $V_n(s)$ is continuous in s.

Theorem 3.2 The optimal policy has the following form: There are increasing numbers $s_1 \leq s_2 \leq \cdots \leq s_n \leq \cdots$ such that if there are n days to go and the present price is s, then one should exercise the option if and only if $s \geq s_n$.

Proof: If the price is s and n days remain, then it follows from Eq. (3.1) that it is optimal to exercise the option if

$$V_n(s) \leq s - c.$$

Let

$$s_n = \min\{s: V_n(s) - s = -c\},$$

where we take s_n to equal ∞ when this set is vacuous. By Lemma 3.1, it can be seen that, for all $s \geq s_n$,

$$V_n(s) - s \leq V_n(s_n) - s_n = -c,$$

thus showing that it is optimal to exercise the option when in state s with n days to go if $s \geq s_n$. That s_n is increasing follows from $V_n(s)$ increasing in n, which is immediate because having additional time cannot decrease one's expected profit. □

Remark: In Problem 5 the reader will be asked to show that it is never optimal to exercise the option before maturity when $\mu_F \geq 0$. Hence Theorem 3.2 is of interest only when the mean drift is negative.

4. Modular Functions and Monotone Policies

Let $g(x, y)$ be a function of two variables, and suppose that for fixed x we are interested in the maximal value attained by g. Let $y(x)$ denote the largest value of y at which the maximum occurs; that is,

$$\max_{y} g(x, y) = g(x, y(x)),$$

---

† In this book the terms *increasing* and *nondecreasing* are used interchangeably.

where $y(x)$ is the largest value of y at which the maximum occurs and where, of course, we are supposing that g is such that $y(x)$ is a well-defined quantity. Under what conditions on g can we assert that $y(x)$ is increasing in x?

To obtain a sufficient condition for $y(x)$ to be increasing, suppose that $x_1 > x_2$ and $y(x_2) = \bar{y}$. One way to guarantee that $y(x_1) \geq \bar{y}$ is to suppose that

$$g(x_1, \bar{y}) \geq g(x_1, y) \qquad \text{for all } y \leq \bar{y}, \tag{4.1}$$

because if Eq. (4.1) is true, then the optimizing y for $g(x_1, y)$ is clearly not less than $\bar{y}$. Because $\bar{y} = y(x_2)$, note that

$$g(x_2, \bar{y}) - g(x_2, y) \geq 0 \qquad \text{for all } y,$$

and thus a sufficient condition for Eq. (4.1) to be valid is that for all $y^* < \bar{y}$,

$$g(x_1, \bar{y}) - g(x_1, y^*) \geq g(x_2, \bar{y}) - g(x_2, y^*).$$

Hence, if

$$g(x_1, y_1) + g(x_2, y_2) \geq g(x_1, y_2) + g(x_2, y_1) \quad \text{for all } x_1 > x_2, \ y_1 > y_2,$$
$$\tag{4.2}$$

then we can conclude that $y(x)$ increases in x.

Definition 4.1   A function g satisfying Eq. (4.2) is called *super-modular*. If the inequality is reversed it is called *submodular*.

If g has mixed partial derivates, then there is a simple characterization of super modularity and submodularity.

Proposition 4.2   If $\partial^2 g(x, y)/\partial x\, \partial y$ exists, then $g(x, y)$ is super-modular if and only if

$$\frac{\partial^2 g(x, y)}{\partial x\, \partial y} \geq 0.$$

Similarly it is submodular if and only if

$$\frac{\partial^2 g(x, y)}{\partial x\, \partial y} \leq 0.$$

Proof: If $\partial^2 g(x, y)/\partial x\, \partial y \geq 0$, then for $x_1 > x_2$ and $y_1 > y_2$,

$$\int_{y_2}^{y_1} \int_{x_2}^{x_1} \frac{\partial^2}{\partial x\, \partial y} g(x, y)\, dx\, dy \geq 0,$$

$$\int_{y_2}^{y_1} \frac{\partial}{\partial y} [g(x_1, y) - g(x_2, y)]\, dy \geq 0,$$

or

$$g(x_1, y_1) - g(x_1, y_2) - g(x_2, y_1) + g(x_2, y_2) \geq 0,$$

which shows that g is supermodular. To go the other way, suppose that g is supermodular. Then for $x_1 > x_2$ and $y_1 > y$,

$$\frac{g(x_1, y_1) - g(x_1, y)}{y_1 - y} \geq \frac{g(x_2, y_1) - g(x_2, y)}{y_1 - y}.$$

Now let y_1 converge to y to obtain

$$\frac{\partial}{\partial y} g(x_1, y) \geq \frac{\partial}{\partial y} g(x_2, y) \qquad \text{when} \quad x_1 \geq x_2,$$

implying that

$$\frac{\partial^2}{\partial x\, \partial y} g(x, y) \geq 0. \quad \square$$

EXAMPLE 4.1 *A Gambling Model with Changing Win Probabilities*
Consider a gambler who is allowed to bet any amount up to his present fortune at each play. He will either win or lose that amount according to a given probability. However, this probability is not fixed but is, in advance of each gamble, chosen at random from some distribution F. On each play the gambler must decide, after the win probability for that play is announced, how much of his fortune to bet. He is allowed to play n times, and his objective is to maximize the expected value of a given function u of his final fortune, where u, called the utility function, is assumed to be increasing in its argument.

The state at any time consists of the gambler's fortune x and the win probability p for that play. Letting $V_n(x, p)$ denote the maximal expected final utility for state (x, p) when only n plays are allowed, then, because the gambler is allowed any bet y between 0 and x, we have the optimality equation

$$V_n(x, p) = \max_{0 \leq y \leq x} \left[p \int_0^1 V_{n-1}(x + y, \alpha) \, dF(\alpha) \right.$$

$$\left. + (1 - p) \int_0^1 V_{n-1}(x - y, \alpha) \, dF(\alpha) \right],$$

with the boundary condition $V_0(x, p) = u(x)$. Let $y_n(x, p)$ denote the largest y that maximizes this optimality equation. Suppose we wanted to prove the rather intuitive result that for fixed n and x the optimal wager is increasing in p. A sufficient condition would be that

$$\frac{\partial^2}{\partial y \, \partial p} \left[p \int_0^1 V_{n-1}(x + y, \alpha) \, dF(\alpha) + (1 - p) \int_0^1 V_{n-1}(x - y, \alpha) \, dF(\alpha) \right] \geq 0,$$

or

$$\int_0^1 \left[\frac{\partial}{\partial y} V_{n-1}(x + y, \alpha) - \frac{\partial}{\partial y} V_{n-1}(x - y, \alpha) \right] dF(\alpha) \geq 0.$$

But this will hold if, for all α,

$$\frac{\partial}{\partial y} \left[V_{n-1}(x + y, \alpha) - V_{n-1}(x - y, \alpha) \right] \geq 0$$

or, equivalently, if, for all α, $V_{n-1}(x + y, \alpha) - V_{n-1}(x - y, \alpha)$ increases in y, which follows if $V_{n-1}(z, \alpha)$ is increasing in z, which immediately follows from the fact that $u(z)$ is increasing.

To show that a given function is supermodular (to conclude a certain monotonicity result about the optimal policy) usually involves more work than in the preceding example, in which this result immediately followed as a consequence of the monotonicity of $V_n(x, \alpha)$ in x. We illustrate by a second example.

EXAMPLE 4.2 *An Optimal Allocation Problem Subject to Penalty Costs*   Suppose we have N stages to construct I successful components sequentially. At each stage we allocate a certain amount of money for the construction of a component. If y is the amount allocated, then the component constructed will be a success with probability $P(y)$, where P is a continuous nondecreasing function satisfying $P(0) = 0$. After each component is constructed, we are informed as to whether or not it is successful. If, at the end of the N stages, we are j components short, then a final penalty cost $C(j)$ is incurred, where $C(j)$ is increasing in j. The problem at each stage is to determine how much money to allocate to minimize the total expected cost (construction cost plus penalty cost) incurred.

We can take as the state the number of successful components still needed. If $V_n(i)$ represents the minimal expected remaining costs when the state is i and n stages remain, then

$$V_n(i) = \min_{y \geq 0} \{y + P(y)V_{n-1}(i-1) + [1 - P(y)]V_{n-1}(i)\}, \quad (4.3)$$

with $V_0(i) = C(i)$.

It follows immediately from the definition of $V_n(i)$ and the monotonicity of $C(i)$ that $V_n(i)$ increases in i and decreases in n. (Verify this.) In many instances it would seem intuitive that "the more we need, the more we would have to invest" and "the more time we have, the less we need invest at each stage." That is, if we let $y_n(i)$ denote the value of y that minimizes Eq. (4.3), then it is somewhat intuitive that

$$y_n(i) \text{ increases in } i \text{ and decreases in } n. \quad (4.4)$$

Let us now attempt to determine conditions on C that make Eq. (4.4) valid. Because we are minimizing rather than maximizing, it follows from modularity that $y_n(i)$ will increase in i if

$$\frac{\partial^2}{\partial i\, \partial y}\{y + P(y)V_{n-1}(i-1) + [1 - P(y)]V_{n-1}(i)\} \leq 0$$

or, equivalently, if

$$P'(y)\frac{\partial}{\partial i}\left[V_{n-1}(i-1) - V_{n-1}(i)\right] \leq 0.$$

Now, $P'(y) \geq 0$ as P is increasing, so

$$y_n(i) \text{ increases in } i \quad \text{if} \quad V_{n-1}(i-1) - V_{n-1}(i) \text{ decreases in } i. \quad (4.5)$$

Similarly,

$$y_n(i) \text{ decreases in } n \quad \text{if} \quad V_{n-1}(i-1) - V_{n-1}(i) \text{ increases in } n. \quad (4.6)$$

Hence, modularity gives us a sufficient condition on the optimal-value function, which ensures the desired monotonicity of the optimal policy. We now show that if $C(i)$ is convex in i, then Eqs. (4.5) and (4.6) hold.

Proposition 4.3   If $C(i + 2) - C(i + 1) \geq C(i + 1) - C(i)$ $(i \geq 0)$, then $y_n(i)$ increases in i and decreases in n.

Proof: Define the following sets of inequalities:

$$A_{i,n}: \quad V_{n+1}(i + 1) - V_{n+1}(i) \leq V_n(i + 1) - V_n(i), \qquad i, n \geq 0,$$

$$B_{i,n}: \qquad V_{n+1}(i) - V_n(i) \leq V_{n+2}(i) - V_{n+1}(i), \qquad i, n \geq 0,$$

$$C_{i,n}: \qquad V_n(i + 1) - V_n(i) \leq V_n(i + 2) - V_n(i + 1), \qquad i, n \geq 0.$$

If we can establish inequalities $C_{i,n}$ and $A_{i,n}$, then, from Eqs. (4.5) and (4.6), Proposition 4.3 will follow. However, it is easier to also establish $B_{i,n}$ at the same time. The proof is by induction on $k = n + i$. Because inequalities $A_{i,n}$, $B_{i,n}$, and $C_{i,n}$ are true when $k = 0$, we can assume that they are true whenever $n + i < k$. Now suppose $n + i = k$. We first show that $A_{i,n}$ is true. Because $A_{i,n}$ is valid when $i = 0$ (because $V_n(i)$ decreases in n), suppose that $i > 0$. Now, for some $\bar{y}$,

$$V_{n+1}(i) = \bar{y} + P(\bar{y})V_n(i - 1) + [1 - P(\bar{y})]V_n(i),$$

thus

$$V_{n+1}(i) - V_n(i) = \bar{y} + P(\bar{y})[V_n(i - 1) - V_n(i)].$$

Also,

$$V_{n+1}(i + 1) \leq \bar{y} + P(\bar{y})V_n(i) + [1 - P(\bar{y})]V_n(i + 1),$$

thus

$$V_{n+1}(i + 1) - V_n(i + 1) \leq \bar{y} + P(\bar{y})[V_n(i) - V_n(i + 1)].$$

Therefore, $A_{i,n}$ will follow if we can show that

$$V_n(i) - V_n(i + 1) \leq V_n(i - 1) - V_n(i).$$

However, this is just inequality $C_{i-1,n}$, which is true when $i + n = k$ by the induction hypothesis.

To prove $B_{i,n}$ we first note that for some $\bar{y}$

$$V_{n+2}(i) = \bar{y} + P(\bar{y})V_{n+1}(i - 1) + [1 - P(\bar{y})]V_{n+1}(i),$$

thus

$$V_{n+2}(i) - V_{n+1}(i) = \bar{y} + P(\bar{y})[V_{n+1}(i - 1) - V_{n+1}(i)].$$

Also,

$$V_{n+1}(i) \leq \bar{y} + P(\bar{y})V_n(i - 1) + [1 - P(\bar{y})]V_n(i),$$

implying that

$$V_{n+1}(i) - V_n(i) \le \bar{y} + P(\bar{y})[V_n(i-1) - V(i)].$$

Therefore $B_{i,n}$ will follow if we can prove that

$$V_n(i-1) - V_n(i) \le V_{n+1}(i-1) - V_{n+1}(i).$$

However, this is just inequality $A_{i-1,n}$, which is consequently true by the induction hypothesis.

To prove $C_{i,n}$, we first note that $B_{i+1,n-1}$ states that

$$2V_n(i+1) \le V_{n-1}(i+1) + V_{n+1}(i+1),$$

and thus $C_{i,n}$ will follow if we can show that

$$V_{n-1}(i+1) + V_{n+1}(i+1) \le V_n(i) + V_n(i+2). \qquad (4.7)$$

Now, for some $\bar{y}$,

$$V_n(i+2) - V_{n-1}(i+1) = \bar{y} + [1 - P(\bar{y})][V_{n-1}(i+2) - V_{n-1}(i+1)].$$

Furthermore,

$$V_{n+1}(i+1) - V_n(i) \le \bar{y} + [1 - P(\bar{y})][V_n(i+1) - V_n(i)].$$

Thus Eq. (4.7) (and $C_{i,n}$) will follow if we can show that

$$V_n(i+1) - V_n(i) \le V_{n-1}(i+2) - V_{n-1}(i+1).$$

Now, from $A_{i,n-1}$ it follows that

$$V_n(i+1) - V_n(i) \le V_{n-1}(i+1) - V_{n-1}(i),$$

and thus it suffices to show that

$$V_{n-1}(i+1) - V_{n-1}(i) \le V_{n-1}(i+2) - V_{n-1}(i+1).$$

However, this is just inequality $C_{i,n-1}$, which is true by the induction hypothesis. □

5. Accepting the Best Offer

Suppose that we are presented with n offers in sequential order. After looking at an offer, we must decide whether to accept it (and terminate the process) or to reject it. Once rejected, an offer is lost. Suppose that the only information we have at any time is the relative rank of the present offer compared with previous ones. The objective is to maximize the probability of selecting the best offer when all $n!$ orderings of the offers are assumed to be equally likely.

We can regard this situation as a sequential decision process in which we say that we are in state i if the ith offer has just been presented and it is the best of the i offers already presented. Letting $V(i)$ denote the best we can do in this position, we find that V satisfies

$$V(i) = \max[P(i), H(i)],$$

where $P(i)$, the probability that the best offer will be realized if the ith is accepted, is given by

$$P(i) = P(\text{offer is best of } n \mid \text{offer is best of first } i)$$

$$= \frac{1/n}{1/i} = \frac{i}{n}$$

and where $H(i)$ represents the best we can do if we reject the ith offer. Hence we have

$$V(i) = \max\left[\frac{i}{n}, H(i)\right], \qquad i = 1, \ldots, n.$$

It is now easy to see that $H(i)$ is just the maximal probability of accepting the best offer when we have rejected the first i offers. But because the situation in which the first i offers have been rejected is clearly at least as good as that in which the first $i + 1$ have been rejected (because the next one can always be rejected), it follows that $H(i)$ is decreasing in i. Because i/n increases and $H(i)$ decreases in i, it follows that for some j

$$\frac{i}{n} \leq H(i) \qquad (i \leq j),$$

$$\frac{i}{n} > H(i) \qquad (i > j).$$

Hence, the optimal policy is of the following form: for some j, $j \leq n - 1$, reject the first j offers and then accept the first candidate offer to appear, where an offer is said to be a candidate if it is of higher value than any of its predecessors.

Letting $P_j(\text{best})$ denote the probability of obtaining the best prize under such a strategy, we have (conditioning on the prize that is accepted.)

$$P_j(\text{best}) = \sum_{i=1}^{n-j} P_j(\text{best} \mid i + j \text{ prize is accepted}) P_j(i + j \text{ accepted}).$$

Now,

$$P_j(\text{best} \mid i + j \text{ accepted}) = P(\text{best of } n \mid \text{best of } i + j)$$

$$= \frac{i + j}{n}.$$

Also,

$$P_j(i + j \text{ accepted}) = P(\text{best of first } j = \text{best of first } i + j - 1,$$

$$i + j = \text{best of first } i + j)$$

$$= P(\text{best of first } j = \text{best of first } i + j - 1)$$

$$\times P(i + j = \text{best of first } i + j)$$

$$= \left(\frac{j}{i + j - 1} \right) \left(\frac{1}{i + j} \right).$$

Hence,

$$P_j(\text{best}) = \frac{j}{n} \sum_{i=1}^{n-j} \frac{1}{i + j - 1}$$

$$= \frac{j}{n} \sum_{k=j}^{n-1} \frac{1}{k}$$

$$\approx \frac{j}{n} \int_{j}^{n-1} \frac{1}{x} \, dx$$

$$= \frac{j}{n} \log\left(\frac{n-1}{j} \right)$$

$$\approx \frac{j}{n} \log\left(\frac{n}{j} \right).$$

Now, if we let $g(x) = (x/n) \log(n/x)$, then

$$g'(x) = \frac{1}{n} \log \frac{n}{x} - \frac{1}{n},$$

so

$$g'(x) = 0 \Rightarrow \log \frac{n}{x} = 1 \Rightarrow x = \frac{n}{e}.$$

Also, because

$$g\left(\frac{n}{e}\right) = \frac{1}{e},$$

we see that the optimal policy is, for n large, approximately to let the fraction $1/e$ of all prizes go by and then accept the first candidate. The probability that this procedure will result in the best prize is roughly $1/e$.

6. A Sequential Allocation Model

Suppose we have D units available for investment. During each of N time periods an opportunity to invest will, independent of the past, occur with probability p. If the opportunity occurs, the investor must decide how much of his remaining wealth to invest. If he invests y, then a return $R(y)$ is earned at the end of the problem. Assuming that both the amount invested and the return become unavailable for future investment, the problem is to decide how much to invest at each opportunity to maximize the expected sum of investment returns.

We assume that $R(y)$ is a nondecreasing, concave function with $R(0) = 0$. Let $V_n(A)$ denote the maximal expected additional profit attainable when there are n time periods to go, A dollars available for investment, and an opportunity is at hand. The function V satisfies the optimality equation

$$V_n(A) = \max_{0 \le y \le A} \left[R(y) + \bar{V}_{n-1}(A - y) \right], \qquad n > 0,$$

$$V_0(A) = 0, \tag{6.1}$$

where

$$\bar{V}_m(A) = \sum_{i=0}^{m} p(1 - p)^i V_{m-i}(A).$$

It should be noted that $\bar{V}_m(A)$ is just the maximal expected additional sum of returns when: A units remain for investment; there are m time periods to go; and it is not yet known if an investment opportunity is available. By conditioning on whether or not an opportunity occurs, we obtain

$$\bar{V}_m(A) = pV_m(A) + (1 - p)\bar{V}_{m-1}(A).$$

We begin by showing that V inherits the concavity property of R.

Lemma 6.1 $V_n(A)$ is a concave function of A.

Proof: The proof is by induction on n. Because $V_1(A) = R(A)$ is concave, assume that $V_i(A)$ is concave in A for $i = 1, \ldots, n - 1$. To prove that V_n is concave we must show that, for $0 < \lambda < 1$,

$$V_n[\lambda A_1 + (1 - \lambda)A_2] \geq \lambda V_n(A_1) + (1 - \lambda)V_n(A_2).$$

Now, for some $y_1 \leq A_1$ and $y_2 \leq A_2$,

$$V_n(A_1) = R(y_1) + \bar{V}_{n-1}(A_1 - y_1),$$
$$V_n(A_2) = R(y_2) + \bar{V}_{n-1}(A_2 - y_2).$$

However, because $\lambda y_1 + (1 - \lambda)y_2 \leq \lambda A_1 + (1 - \lambda)A_2$, it follows from the optimality equation (6.1) that

$$\begin{aligned}
V_n[\lambda A_1 + (1 - \lambda)A_2] &\geq R[\lambda y_1 + (1 - \lambda)y_2] \\
&\quad + \bar{V}_{n-1}[\lambda(A_1 - y_1) + (1 - \lambda)(A_2 - y_2)] \\
&\geq \lambda R(y_1) + (1 - \lambda)R(y_2) + \lambda \bar{V}_{n-1}(A_1 - y_1) \\
&\quad + (1 - \lambda)\bar{V}_{n-1}(A_2 - y_2) \\
&= \lambda V_n(A_1) + (1 - \lambda)V_n(A_2),
\end{aligned}$$

where the second inequality follows from the concavity of R and the concavity of $\bar{V}_{n-1}$, the latter of which follows from the induction hypothesis. Thus $V_n(A)$ is concave in A. $\square$

Let us now define $y_n(A)$ to be the value of y (or the smallest value of y if there is more than one) that maximizes the right-hand side of the optimality equation (6.1). That is, $y_n(A)$ is the optimal amount to invest when the available investment capital equals A, there are n time periods remaining, and an opportunity to invest is at hand. The next result yields the structure of the optimal policy.

Theorem 6.2

(i) $y_n(A)$ is a nondecreasing function of A,
(ii) $y_n(A)$ is a nonincreasing function of n.

Proof: To prove (i) let $\bar{y} = y_n(A)$. Then, by its definition, it follows that for $y < \bar{y}$

$$R(\bar{y}) + \overline{V}_{n-1}(A - \bar{y}) > R(y) + \overline{V}_{n-1}(A - y). \tag{6.2}$$

For $\epsilon > 0$, we shall show that $y_n(A + \epsilon) \geq \bar{y}$ by proving that, whenever $y < \bar{y}$,

$$R(\bar{y}) + \overline{V}_{n-1}(A + \epsilon - \bar{y}) \geq R(y) + \overline{V}_{n-1}(A + \epsilon - y).$$

Now, from Eq. (6.2), this will be proven if we can show that, whenever $y < \bar{y}$,

$$\overline{V}_{n-1}(A - y) - \overline{V}_{n-1}(A - \bar{y}) \geq \overline{V}_{n-1}(A + \epsilon - y) - \overline{V}_{n-1}(A + \epsilon - \bar{y}).$$

However, this follows from the concavity of V_n, which implies the concavity of $\overline{V}_{n-1}$. Thus (i) is proven.

To prove that $y_n(A) \geq y_{n+1}(A)$, let $\bar{y} = y_n(A)$. If $\bar{y} = A$, then the result is immediate; so suppose that $\bar{y} < A$. Fix $\epsilon < A - \bar{y}$ and let $y^* = y_n(A - \bar{y} - \epsilon)$. Note that by part (i) of Theorem 6.2, it follows that $y^* \leq \bar{y}$. Now,

$$
\begin{aligned}
R(\bar{y} + \epsilon) + V_n(A - \bar{y} - \epsilon) &= R(\bar{y} + \epsilon) + R(y^*) + \overline{V}_{n-1}(A - y - \epsilon - y^*) \\
&\leq R(\bar{y}) + R(y^* + \epsilon) + \overline{V}_{n-1}(A - \bar{y} - \epsilon - y^*) \\
&\leq R(\bar{y}) + V_n(A - \bar{y}), \tag{6.3}
\end{aligned}
$$

where the first inequality follows from the concavity of R and the fact that $y^* \leq \bar{y}$. By using the relationship

$$\overline{V}_n(x) = p V_n(x) + (1 - p)\overline{V}_{n-1}(x),$$

we have

$$
\begin{aligned}
&R(\bar{y} + \epsilon) + \overline{V}_n(A - \bar{y} - \epsilon) - R(\bar{y}) - \overline{V}_n(A - \bar{y}) \\
&\quad = p\big[R(\bar{y} + \epsilon) + V_n(A - \bar{y} - \epsilon) - R(\bar{y}) - V_n(A - \bar{y})\big] \\
&\quad\quad + (1 - p)\big[R(\bar{y} + \epsilon) + \overline{V}_{n-1}(A - \bar{y} - \epsilon) - R(\bar{y}) - \overline{V}_{n-1}(A - \bar{y})\big].
\end{aligned}
$$

Now, the first term in square brackets is nonpositive by Eq. (6.3), and the second is nonpositive because $\bar{y} = y_n(A)$. Therefore

$$R(\bar{y} + \epsilon) + \overline{V}_n(A - \bar{y} - \epsilon) \leq R(\bar{y}) + \overline{V}_n(A - \bar{y}),$$

implying that

$$y_{n+1}(A) \leq \bar{y} = y_n(A),$$

and the proof is complete. □

Thus, when the return from an investment is a concave function of the amount invested, then "the more one has, the more one should invest," and "the more time one has, the less one should invest." These results, which depend on R being concave, are quite intuitive, for suppose that one knew that there would be k opportunities to invest. Then, from concavity (see Problem 7), it would be optimal to invest A/k at each opportunity. Hence it is quite intuitive that as the number of opportunities increases stochastically the amount invested decreases and as A increases the amount invested increases.

7. The Interchange Argument in Sequencing

In some problems, a policy consists of a sequence of decisions that is fixed at time zero. Often in such problems, a valuable technique is to consider an arbitrary sequence and then to see what happens when two adjacent decisions are interchanged. We illustrate by an example.

Consider an individual who is presented with n tasks. The ith task has value V_i and has a known probability P_i that it will be successfully performed. Once the individual fails at a task, he is no longer allowed to try any further ones. The problem of interest is to determine which of the $n!$ possible task orderings results in the largest expected sum of values obtained before a failure occurs.

To solve this problem, let us consider any ordering and see how the expected return is changed when we interchange two adjacent tasks. Let us start by interchanging the first two tasks. That is, let $O_1 = (i, j, i_3, \ldots, i_n)$ and $O_2 = (j, i, i_3, \ldots, i_n)$. Now,

$$E(\text{return under } O_1) = P_i V_i + P_i P_j V_j + P_i P_j E(\text{return from } i_3, \ldots, i_n),$$

whereas

$$E(\text{return under } O_2) = P_j V_j + P_j P_i V_i + P_j P_i E(\text{return from } i_3, \ldots, i_n).$$

Hence,

$$E(\text{return under } O_1) \leq E(\text{return under } O_2) \Leftrightarrow \frac{P_i V_i}{1 - P_i} \leq \frac{P_j V_j}{1 - P_j}. \quad (7.1)$$

Now suppose we interchange any two adjacent elements, that is, suppose $O_3 = (i_1, \ldots, i_k, i, j, \ldots, i_n)$ and $O_4 = (i_1, \ldots, i_k, j, i, \ldots, i_n)$. Then

$$E(\text{return from } O_3) = E[\text{return from } (i_1, \ldots, i_k)]$$

$$+ \prod_{l=1}^{k} P_{i_l} E[\text{return from } (i, j, \ldots, i_n)],$$

and

$$E(\text{return from } O_4) = E[\text{return from } (i_1, \ldots, i_k)]$$

$$+ \prod_{l=1}^{k} P_{i_l} E[\text{return from } (j, i, \ldots, i_n)],$$

so from Eq. (7.1) we see that interchanging j with i leads to an improvement if and only if

$$\frac{P_i V_i}{1 - P_i} \leq \frac{P_j V_j}{1 - P_j}.$$

But this shows that the optimal sequence is in decreasing order of $P_j V_j / (1 - P_j)$. Suppose $P_1 V_1 / (1 - P_1) \geq P_2 V_2 / (1 - P_2) \geq \cdots \geq P_n V_n / (1 - P_n)$. Then consider any ordering where 1 is not first. By successive interchanges we obtain a better order in which 1 is first. If task 2 is not now in second position, we again make successive interchanges, each leading to an increased expected return, until it is in second position, and so on.

EXAMPLE 7.1a *Job Scheduling Model* An individual has n jobs, that he must perform in a sequential order. The ith job requires a random time X_i for its execution, where it is assumed that $X_1, \ldots, X_n$ are independent. In addition, if the ith job is completed at time t, the individual earns $\alpha^t R_i$, $i = 1, \ldots, n$, where the quantity α $(0 > \alpha > 1)$ is called the discount factor and is used to indicate that a fixed amount of money to be earned at time t in the future loses value according to the factor α^t. The objective is to choose the ordering of jobs to maximize the expected total return.

Consider any ordering of jobs $O_1 = (i_1, \ldots, i_k, i, j, \ldots i_n)$, and let us see what happens if we interchange the two adjacent jobs i and j. That is, consider also $O_2 = (i_1, \ldots, i_k, j, i, \ldots, i_n)$. Conditioning on the time it takes to complete work on jobs $i_1, \ldots, i_k$, we find that the conditional expected difference in return from the two orderings is

$$E\left[\text{return from } O_1 - \text{return from } O_2 \mid \sum_{l=1}^{k} X_{i_l} = t\right]$$

$$= E\left[\alpha^{t+X_i}R_i + \alpha^{t+X_i+X_j}R_j - \alpha^{t+X_j}R_j - \alpha^{t+X_j+X_i}R_i\right]$$

$$= \alpha^t\left[R_iE(\alpha^{X_i}) + R_jE(\alpha^{X_i})E(\alpha^{X_j}) - R_jE(\alpha^{X_j}) - R_iE(\alpha^{X_i})E(\alpha^{X_j})\right],$$

where the last equality makes use of the assumed independence of the X_k $(k = 1, \ldots, n)$. Hence we see that

$$E(\text{return from } O_1) \leq E(\text{return from } O_2) \Leftrightarrow \frac{R_iE(\alpha^{X_i})}{1 - E(\alpha^{X_i})} \leq \frac{R_jE(\alpha^{X_j})}{1 - E(\alpha^{X_j})}.$$

However, from this we can conclude that the optimal policy is to schedule jobs in decreasing order of $[R_iE(\alpha^{X_i})]/[1 - E(\alpha^{X_i})]$. (We leave the details as an exercise.)

EXAMPLE 7.1b *Goofspiel*   The game of pure strategy, sometimes called Goofspiel, is played by two players using a normal deck of cards as follows. The 13 clubs are first taken out of the deck, and of the remaining 39 cards, the 13 hearts are given to Player I, the 13 diamonds to Player II, and the 13 spades are placed face down in the center. The spades are shuffled and one is turned face up. At this point, the two players choose one of their cards and then simultaneously discard them. The one who discards the higher card (ace being low, king high) wins from the other an amount equal to the value of the upturned spade (ace = 1, king = 13). If both players discard the same card, then neither wins. The three cards are then thrown away, a new space upturned and the game continues. After 13 plays, there are no remaining cards and the game ends.

Consider this game under the assumption that Player II discards his cards in a completely random manner. Given this information, we show that the best thing for Player I to do is always to match the upturned spade, that is, if the upturned card is an ace, then Player I should play his ace, etc.

Let us first generalize our game. Suppose that Player I has N cards having values $V_1, V_2, \ldots, V_N$, where $V_1 \leq V_2 \leq \cdots \leq V_N$; Player II has N cards having values $Y_1, Y_2, \ldots, Y_N$, where $Y_1 \leq Y_2 \leq \cdots \leq Y_N$; and the N cards in the middle have values $P_1, P_2, \ldots, P_N$, where $P_1 \leq$

$P_2 \leq \cdots \leq P_N$. The game is played as before. One of the center cards is turned face up. The players then simultaneously discard and whoever has the higher-value card wins from the other an amount equal to the value of the middle card. These three cards are then thrown away and the play continues until there are no cards left.

Theorem 7.1   If Player II discards in a completely random manner, then the strategy maximizing Player I's expected winning is the one that discards the card having value V_i whenever the upturned middle card has value P_i, $i = 1, 2, \ldots, N$.

Proof: The proof is by induction on N. The theorem is trivially true for $N = 1$, so assume it for $N - 1$. Suppose now that for the *N-card* problem the initial upturned card has value P_j and consider any strategy that calls for Player I to play V_i, where $i < j$. After this first discard, Player I has cards $V_1, \ldots, V_{i-1}, V_{i+1}, \ldots, V_j, \ldots, V_N$ and the center has cards $P_1, \ldots, P_i, \ldots, P_{j-1}, P_{j+1}, \ldots, P_N$. Hence, from the induction hypothesis, it follows that, if the initial upturned card has value P_j, then among those strategies that call for Player I to play V_i, the best is the one that plays

$$
\begin{aligned}
V_k & \quad \text{on} \quad P_k, \quad k = 1, \ldots, i - 1, \\
V_i & \quad \text{on} \quad P_j, \\
V_{k+1} & \quad \text{on} \quad P_k, \quad k = i, \ldots, j - 1, \\
V_k & \quad \text{on} \quad P_k, \quad k = j + 1, \ldots, N.
\end{aligned}
\tag{7.2}
$$

Compare this, however, with the strategy that is the same as that in Eq. (7.2), with the exception that it uses

$$
\begin{aligned}
V_j & \quad \text{on} \quad P_j, \\
V_i & \quad \text{on} \quad P_{j-1}.
\end{aligned}
\tag{7.3}
$$

That is, the strategies of Eqs. (7.2) and (7.3) are identical except that Eq. (7.2) uses

$$
\begin{aligned}
V_i & \quad \text{on} \quad P_j, \\
V_j & \quad \text{on} \quad P_{j-1}
\end{aligned}
$$

in contrast to Eq. (7.3). The expected payoff to Player I for these two plays is, under the strategy of Eq. (7.2),

$$P_j\left[\frac{2(\text{number } k: Y_k < V_i)}{N} - 1\right] + P_{j-1}\left[\frac{2(\text{number } k: Y_k < V_j)}{N} - 1\right],$$

whereas under the strategy of Eq. (7.3), it is the same expression with P_j and P_{j-1} interchanged. Because

$$P_j \geq P_{j-1}, \qquad (\text{number } k: Y_k < V_j) \geq (\text{number } k: Y_k < V_i),$$

it follows that the expected payoff is at least as good for the strategy of Eq. (7.3) as for that of Eq. (7.2), and hence there is a strategy that initially plays V_j that is at least as good as the one playing V_i when $i < j$. A similar result may be shown when $i > j$, and hence by the induction hypothesis the strategy that always matches the rank of the upturned card is optimal.

Problems

1. A deck of 52 cards is to be turned over one at a time. Before each card is turned we are given the opportunity to say whether or not it will be the ace of spades. We are allowed to say that a card will be the ace of spades only once and our objective is to maximize the probability of being correct.
   (a) Set this up as a dynamic programming problem.
   (b) What is an optimal policy?
   (c) What would an optimal policy be if we were given n chances to select the ace of spades?
   (d) What would an optimal policy be if our objective were to select any of the 13 spades (and not just the ace)?

2. Prove that the policy that chooses the action a that maximizes $R(i, a) + \sum_j P_{ij}(a)V_{n-1}(j)$, when there are n time periods to go and the present state is i, is an optimal policy. That is, show that under this policy the expected return, for an N-stage problem starting in state i, is $V_N(i)$.

3. For the model of Section 2, show that, when $p \leq \frac{1}{2}$, $V_n(x) = x$, and the optimal strategy is to always bet 0. In fact, prove this when the

objective is to maximize the expected value of $u(X)$, where X is the final fortune and u is a concave function.

4. For the model of Section 3, show that $V_n(s)$ is a continuous function of s.

5. For the stock-option model of Section 3, show that when $\mu_F \geq 0$, $s_n = \infty$ for $n \geq 1$.

6. Requisitions arrive daily at a data processing center. With probability $P_j(j \geq 0)$, there will be j requisitions in a day. Right after the daily requisitions have arrived, a decision has to be made as to whether or not to process. The cost of processing a batch of any size is K and the processing time is negligible. A penalty cost of c per day per requisition is incurred while requisitions await processing. All requisitions must be processed by time N, and the objective is to minimize the expected sum of costs. (Note that if we decide to process, then all of the requisitions still awaiting processing are simultaneously processed at a fixed cost K.)
    (a) Write the optimality equation.
    (b) Show that the optimal policy has the following form: it processes at time n if and only if the number of requisitions accumulated just after time n is greater than or equal to some value s_n. Show how to determine these critical numbers $s_1, \ldots, s_N$.
    (c) Is the sequence $s_1, \ldots, s_N$ monotone?

7. Let $\{f_j(x), j = 1, 2, \ldots, n\}$ be a set of n functions, and consider the problem of maximizing $\sum_{j=1}^{n} f_j(x_j)$, where the maximum is over all nonnegative vectors $\mathbf{x} = (x_1, \ldots, x_n)$ such that $\sum_{j=1}^{n} x_j = x$.
    (a) Set this up as a dynamic programming problem and write the optimality equation. Now suppose

    $$f_j(x) = f(x) \qquad (j = 1, \ldots, n).$$

    Determine the optimal solution when f is an increasing and
    (b) convex function,
    (c) concave function.

8. We have at most n attempts to construct a successful system. If x units are invested in such a construction, then the resultant system will be successful with probability $P(x)$, where $P(x)$ is increasing and $P(0) = 0$. We have a total income of A units, and

the problem of interest is to determine how much to invest at each attempt to maximize the probability of obtaining a successful system. Let x_i denote the optimal amount to allocate on the ith opportunity ($i = 1, \ldots, n$).

(a) If $\log[1 - P(x)]$ is convex, show that $x_1 = x_2 = \cdots = x_n = A/n$.

(b) If $\log[1 - P(x)]$ is concave, show that $x_1 = A$, $x_2 = \cdots = x_n = 0$.

9. At the beginning of each of N time periods, an individual must decide how much of his fortune to consume and how much to save. If his fortune is x and he consumes y ($0 \le y \le x$), then he obtains a utility $u(y)$, and the amount $x - y$ is saved for the next period. In addition, at the beginning of each period, before the decision as to how much to consume is made, the individual receives an additional random amount (which can be interpreted as his wages from the previous period's work), which has distribution F. The objective is to maximize the expected sum of utilities obtained over the first N time periods. Set this up as a dynamic programming problem and write the optimality equation.

10. Consider a set of n jobs, where job i takes a random time X_i, having distribution F_i, to be performed. If job i is completed at time t, a reward $R_i(t)$ ($i = 1, \ldots, n$) is earned. The objective is to maximize the total expected return (that is, the expected sum of the rewards) when there is a single worker who must perform the jobs in some sequential order.

(a) Assuming that the worker can decide after each job completion which job to attempt next, argue that the state space is the set of pairs (S, t), where S is a subset of $\{1, 2, \ldots, n\}$ and $t \ge 0$. Explain what it means for the state to be (S, t). Let $V(S, t)$ denote the maximal expected return from time t onward, when the state is (S, t). Write a functional equation for $V(S, t)$.

(b) Suppose now that the order in which the jobs are to be performed must be chosen at time 0 (so there are only $n!$ possible policies). Letting $V(S)$ denote the maximal return when S, $S \subset \{1, 2, \ldots, n\}$, is the original set of jobs, argue that

$$V(S) = \max_{i \in S}\{V[\{S - i\}] + E[R_i(\sum_{j \in S} X_j)]\},$$

where $\{S - i\}$ is the set consisting of all elements in S except for i. Explain how this recursion can be used to determine the optimal policy.

11. An urn initially contains n white and m black balls. Balls are selected at random, without replacement, and the decision maker wins one unit when a white ball is selected and loses one unit when a black ball is selected. At any point in time, the decision maker is allowed to quit playing. The objective is to maximize expected winnings.
   (a) Set this up as a dynamic programming problem with state space $\{(k, r) : k \le n, r \le m\}$. Interpret the states.
   (b) Write the optimality equation and explain how the problem can be solved.
   (c) Suppose it is optimal to play in state (k, r). Prove or give a counterexample to the following claims:
       (i)   it is optimal to play in state $(k + 1, r)$,
       (ii) it is optimal to play in state $(k - 1, r)$,
       (iii) it is optimal to play in state $(k, r + 1)$,
       (iv) it is optimal to play in state $(k, r - 1)$.
   (d) Repeat part (c) when the word "play" is changed to "stop" throughout.

12. N offers are to be received sequentially in time, each offer being a value chosen from a known distribution F. After observation each offer must either be accepted or rejected.
   (a) Suppose that the objective is to maximize the expected value of the accepted offer. Set this up as a dynamic programming problem and show that there are numbers $a_1 \le a_2 \le \cdots \le a_N$ such that, if there are n offers remaining, then the present one should be accepted if its value is at least a_n. What is a_1?
   (b) Suppose now that the objective is to maximize the probability of obtaining the best offer. Show that there are numbers $a_1 \le a_2 \le \cdots$ such that, if there are n offers to go, then the present one should be accepted if it is the best one yet seen and its value is at least a_n. What is a_1?

13. In Example 4.1, show that, for $u(x) = \log x$,

$$y_n(x, p) = \begin{cases} (2p - 1)x, & p \ge \frac{1}{2} \\ 0, & p < \frac{1}{2}. \end{cases}$$

14. In Example 4.2, suppose that $C(j) = jA$. Show that $y_n(i)$, the value

of y that minimizes the right side of Eq. (4.3) when $V_0(i) = iA$, is a nondecreasing function of A.

15. For the investment model of Section 6, do the following:
   (a) Show that

   $$\overline{V}_n(A) = pV_n(A) + (1 - p)\overline{V}_{n-1}(A).$$

   (b) Show, if $R(y)$ is a nondecreasing continuous function, that $V_n(A)$ is
      (i) a continuous, nondecreasing function of A and
      (ii) nondecreasing in n if $R(x) \geq 0$.
   (c) If $R(y)$ is convex, show that it is optimal to invest everything when an opportunity presents itself.

16. For the investment model of Section 6, suppose that $R(y)$ is a nondecreasing concave function. Let $y_n(A)$ denote the optimal amount to invest when an opportunity is present with n time periods to go and A units available for investment.
   (a) Prove $y_n(A + B) \leq B + y_n(A)$.
      Interpret this as meaning "the more you have, the more you conserve."
   (b) Use the results of Problem 7 to show

   $$V_n(A) \leq \sum_{i=0}^{n-1} \binom{n-1}{i} p^i (1 - p)^{n-1-i}(i + 1)R\left(\frac{A}{i + 1}\right).$$

   (c) If $R(y) = \log y$, show that

   $$y_n(A) = \frac{A}{1 + (n - 1)p}.$$

   Hint: First prove by induction that

   $$\overline{V}_n(A) = \overline{V}_{n-1}(1) + np \log A.$$

17. For the sequential allocation model of Section 6, prove that

   (a) $\displaystyle \overline{V}_n(A) \leq \sum_{i=1}^{n} \binom{n}{i} p^i (1 - p)^{n-i} iR(A/i) \leq npR(A/np).$

   Hint: For the first inequality, make use of the results of Problem 7, and for the second, recall Jensen's inequality, which states that for a concave function f

   $$E[f(X)] \leq f(E[X])$$

(b) $\lim_{n\to\infty} \bar{V}_n(A)/npR(A/np) = 1.$

(c)   Use (b) to argue that, for n large,

$$y_n(A) \simeq \frac{A}{1 + (n-1)p}.$$

18.   There are two gold mines, I and II, that initially contain, respectively, A and B tons of gold. There is a single gold-mining machine. Each time the machine is used at mine I (II) there is a probability P_1 (P_2) that it will mine the fraction r_1 (r_2) of the remaining gold at the mine and remain in working order and a probability $1 - P_1$ ($1 - P_2$) that it will mine no gold and be damaged beyond repair. The objective is to maximize the expected amount of gold mined before the machine breaks down. Determine the optimal policy.

19.   In Goofspiel, determine the expected winnings of Player I if Player II is known to discard cards in a completely random fashion.

Notes and References

The gambling model of Section 2 is due to Kelly [8], and the stock-option model of Section 3 is a special case of models considered by Taylor [11]. The use of modularity in determining the structure of optimal policies has been championed by Topkis [12]. Example 4.1 is from Subelman [10] and Example 4.2 is from Derman *et al.* [3]. Section 5 is based on Gilbert and Mosteller [7] and Section 6 on Derman *et al.* [4]. Example 7.1b is from Ross [9]. The texts by Bellman [1], Dreyfus and Law [5], and White [13] are introductory dynamic programming, whereas those by Bertsekas [2] and Dynkin [6] are at a higher mathematical level.

1. Bellman, R., *Dynamic Programming*. Princeton University Press, Englewood Cliffs, N.J., 1957.

2. Bertsekas, D., *Dynamic Programming and Stochastic Control*. Academic Press, New York, 1976.

3. Derman, C., Lieberman, G. J., and Ross, S., "Optimal system allocation with penalty costs," *Management Sci.* **23**, 399–404, 1976.

4. Derman, C., Lieberman, G. J., and Ross, S., "A stochastic sequential allocation model," *Operations Res.* **23**, 1120–1130, November 1975.

5. Dreyfus, S., and Law, A., *The Art and Theory of Dynamic Programming*. Academic Press, New York, 1976.

6. Dynkin, E. B., *Controlled Markov Processes*. Springer-Verlag, New York, 1979.

7. Gilbert, J., and Mosteller, F., "Recognizing the maximum of a sequence," *J. Am. Stat. Assoc.* **61,** 35–73, 1966.

8. Kelly, J., "A new interpretation of information rate," *Bell System Tech. J.* **35,** 917–926, 1956.

9. Ross, S. M., "Goofspiel—the game of pure strategy," *J. Appl. Prob.* **8,** 621–625, 1971.

10. Subelman, E., "The dependence of betting strategies on the probability of winning," *J. Appl. Prob.* **16,** 855–866, 1971.

11. Taylor, H., "Evaluating a call option and optimal timing strategy in the stock market," *Management Sci.* **12,** 111–120, 1967.

12. Topkins, D. M., "Minimizing a submodular function on a lattice," *Operations Res.* **26,** 305–321, 1978.

13. White, D. J., *Dynamic Programming*. Holden–Day, San Francisco, 1969

Discounted Dynamic Programming

1. Introduction

Consider a process that is observed at time points $n = 0, 1, 2, \ldots$ to be in one of a number of possible states. The set of possible states is assumed to be countable and will be labeled by the nonnegative integers $0, 1, 2, \ldots$. After observing the state of the process, an action must be chosen, and we let A (assumed finite) denote the set of all possible actions.

If the process is in state i at time n and action a is chosen, then, independent of the past, two things occur:

(i)   We receive an expected reward $R(i, a)$.

(ii)   The next state of the system is chosen according to the transition probabilities $P_{ij}(a)$.

If we let X_n denote the state of the process at time n and a_n the action chosen at that time, then assumption (ii) is equivalent to stating that

$$P\{X_{n+1} = j | X_0, a_0, X_1, a_1, \ldots, X_n = i, a_n = a\} = P_{ij}(a).$$

Thus both the rewards and the transition probabilities are functions only of the last state and the subsequent action. Furthermore, we suppose that the costs are bounded, and we let B be such that $|R(i, a)| < B$ for all i and a.

To choose actions we must follow some policy. We place no restrictions on the class of allowable policies, and we therefore define a *policy* to be any rule for choosing actions. Thus the action chosen by

29

a policy, for instance, may depend on the history of the process up to that point or it may be randomized in the sense that it chooses action a with some probability P_a, $a \in A$.

An important subclass of the class of all policies is the class of stationary policies. Here, a policy is said to be *stationary* if it is non-randomized and the action it chooses at time t only depends on the state of the process at time t. In other words, a stationary policy is a function f mapping the state space into the action space, with the interpretation that for each state i, $f(i)$ denotes the action the policy chooses when in state i. It follows that if a stationary policy f is employed, then the sequence of states $\{X_n, n = 0, 1, 2, \ldots\}$ forms a Markov chain with transition probabilities $P_{ij} = P_{ij}(f(i))$; and it is for this reason that the process is called a Markov decision process.

To determine policies that are in some sense optimal, we first need to decide on an optimality criterion. In this chapter we use the total expected discounted return as our criterion. This criterion assumes a discount factor α, $0 < \alpha < 1$, and, among all policies π, attempts to maximize

$$V_\pi(i) = E_\pi \left[\sum_{n=0}^{\infty} R(X_n, a_n)\alpha^n \middle| X_0 = i \right], \tag{1.1}$$

where E_π represents the conditional expectation, given that policy π is employed. Because $R(X_n, a_n)$ is just the reward earned at time n, it follows that $V_\pi(i)$ represents the expected total discounted return earned when the policy π is employed and the initial state is i. [Note that (1.1) is well defined because rewards are bounded and $\alpha < 1$, which implies that $|V_\pi(i)| < B/(1 - \alpha)$.]

2. The Optimality Equation and Optimal Policy

The use of a discount factor is economically motivated by the fact that a reward to be earned in the future is less valuable than one earned today. Let

$$V(i) = \sup_\pi V_\pi(i).$$

A policy π^* is said to be α-*optimal* if

$$V_{\pi^*}(i) = V(i) \qquad \text{for all } i \geq 0.$$

Hence, a policy is α-optimal if its expected α-discounted return is maximal for every initial state.

The following theorem yields a functional equation satisfied by the optimal value function V.

Theorem 2.1   The Optimality Equation

$$V(i) = \max_a \Big[R(i, a) + \alpha \sum_j P_{ij}(a)V(j) \Big], \qquad i \geq 0. \qquad (2.1)$$

Proof: Let π be any arbitrary policy, and suppose that π chooses action a at time 0 with probability P_a, $a \in A$. Then,

$$V_\pi(i) = \sum_{a \in A} P_a \Big[R(i, a) + \sum_j P_{ij}(a)W_\pi(j) \Big],$$

where $W_\pi(j)$ represents the expected discounted return from time 1 onward, given that policy π is being used and that the state at time 1 is j. However, if the state at time 1 is j, the situation at this time is the same as if the process had started in state j, with the exception that all returns are now multiplied by α. Hence,

$$W_\pi(j) \leq \alpha V(j),$$

and thus

$$\begin{aligned}
V_\pi(i) &\leq \sum_{a \in A} P_a \Big[R(i, a) + \alpha \sum_j P_{ij}(a)V(j) \Big] \\
&\leq \sum_{a \in A} P_a \max_a \Big[R(i, a) + \alpha \sum_j P_{ij}(a)V(j) \Big] \\
&= \max_a \Big[R(i, a) + \alpha \sum_j P_{ij}(a)V(j) \Big]. \qquad (2.2)
\end{aligned}$$

Because π is arbitrary, (2.2) implies that

$$V(i) \leq \max_a \Big[R(i, a) + \alpha \sum_j P_{ij}(a)V(j) \Big]. \qquad (2.3)$$

To go the other way, let a_0 be such that

$$R(i, a_0) + \alpha \sum_j P_{ij}(a_0)V(j) = \max_a \Big[R(i, a) + \alpha \sum_j P_{ij}(a)V(j) \Big]. \quad (2.4)$$

Let π be the policy that chooses a_0 at time 0 and, if the next state is j, then views the process as originating in state j, following a policy π_j such that $V_{\pi_j}(j) \geq V(j) - \epsilon$. Hence,

$$V_\pi(i) = R(i, a_0) + \alpha \sum_j P_{ij}(a_0) V_{\pi_j}(j)$$

$$\geq R(i, a_0) + \alpha \sum_j P_{ij}(a_0) V(j) - \alpha\epsilon,$$

which, because $V(i) \geq V_\pi(i)$, implies that

$$V(i) \geq R(i, a_0) + \alpha \sum_j P_{ij}(a_0) V(j) - \alpha\epsilon.$$

Hence, from (2.4) we obtain

$$V(i) \geq \max_a \left[R(i, a) + \alpha \sum_j P_{ij}(a) V(j) \right] - \alpha\epsilon, \qquad (2.5)$$

and the result follows from (2.3) and (2.5) because ϵ is arbitrary. □

We are now ready to prove the important result that the policy determined by the optimality equation is optimal.

Theorem 2.2   Let f be the stationary policy that, when the process is in state i, selects the action (or an action) maximizing the right side of (2.1), that is, $f(i)$ is such that

$$R(i, f(i)) + \alpha \sum_j P_{ij}(f(i)) V(j) = \max_a \left[R(i, a) + \alpha \sum_j P_{ij}(a) V(j) \right], \quad i \geq 0.$$

Then

$$V_f(i) = V(i) \qquad \text{for all } i \geq 0,$$

and hence f is α-optimal.

Proof: Because

$$V(i) = \max_a \left[R(i, a) + \alpha \sum_j P_{ij}(a) V(j) \right]$$

$$= R(i, f(i)) + \alpha \sum_j P_{ij}(f(i)) V(j), \qquad (2.6)$$

we see that V is equal to the expected discounted return of a two-stage problem in which we use f for the first stage and then receive a terminal reward $V(j)$ (if we end in state j). But because this terminal reward has the same value as using f for another stage and then receiving the terminal reward V, we see that V is equal to the expected reward of a three-stage problem in which we use f for two stages and then receive the terminal reward V. Continuing this argument shows that

$$V(i) = E(n\text{-stage return under } f \,|\, X_0 = i) + \alpha^n E(V(X_n) \,|\, X_0 = i).$$

Letting $n \to \infty$, we obtain, using $V(j) < B/(1 - \alpha)$ and $0 < \alpha < 1$,

$$V(i) = V_f(i),$$

which proves the theorem. □

Technical Remark: The preceding proof can be stated more formally as follows: For any stationary policy g, define the operator T_g mapping bounded functions on the state space into itself in the following manner. For any bounded function $u(i)$ $(i = 0, 1, \ldots)$, $T_g u$ is defined as that function whose value at i is given by

$$(T_g u)(i) = R(i, g(i)) + \alpha \sum_j P_{ij}(g(i)) u(j).$$

Thus $T_g u$ evaluated at i represents the expected discounted return if the initial state is i and we employ g for one stage and are then terminated with a final return $u(j)$ (if the final state is j).

It is easy to show that for bounded functions u and v

(i) $u \le v \Rightarrow T_g u \le T_g v,$
(ii) $T_g^n u \to V_g$ as $n \to \infty,$

where $T_g^1 u = T_g u$, $T_g^n u = T_g(T_g^{n-1} u)$, $n \ge 1$.

If f is the policy chosen by the optimality equation, we have by (2.1) that

$$T_f V = V,$$

implying that

$$T_f^n V = T_f^{n-1}(T_f V) = T_f^{n-1} V = \cdots = V,$$

and, letting $n \to \infty$,

$$V_f = V. \quad \square$$

The following proposition shows that V is the unique bounded solution of the optimality equation.

Proposition 2.3 V is the unique bounded solution of the optimality equation (2.1).

Proof: Suppose that $u(i)$, $i \geq 0$, is a bounded function that satisfies the optimality equation

$$u(i) = \max_a \left[R(i, a) + \alpha \sum_j P_{ij}(a)u(j) \right], \qquad i \geq 0.$$

For fixed i let $\bar{a}$ be such that

$$u(i) = R(i, \bar{a}) + \alpha \sum_j P_{ij}(\bar{a})u(j).$$

Hence, because V satisfies the optimality equation, we have

$$u(i) - V(i) = R(i, \bar{a}) + \alpha \sum_j P_{ij}(\bar{a})u(j) - \max_a \left[R(i, a) + \alpha \sum_j P_{ij}(a)V(j) \right]$$

$$\leq \alpha \sum_j P_{ij}(\bar{a})[u(j) - V(j)]$$

$$\leq \alpha \sum_j P_{ij}(\bar{a})|u(j) - V(j)|$$

$$\leq \alpha \sum_j P_{ij}(\bar{a}) \sup_j |u(j) - V(j)|$$

$$= \alpha \sup_j |u(j) - V(j)|.$$

By reversing the roles of u and V we can similarly conclude that

$$V(i) - u(i) \leq \alpha \sup_j |V(j) - u(j)|.$$

Therefore

$$|V(i) - u(i)| \leq \alpha \sup_j |V(j) - u(j)|,$$

so

$$\sup_i |V(i) - u(i)| \leq \alpha \sup_j |V(j) - u(j)|,$$

implying (because $\alpha < 1$) that

$$\sup_j |V(j) - u(j)| = 0. \qquad \square$$

The following will also be needed later.

Proposition 2.4 For any stationary policy g, V_g is the unique solution of

$$V_g(i) = R(i, g(i)) + \alpha \sum_j P_{ij}(g(i))V_g(j). \tag{2.7}$$

Proof: It is immediate that V_g satisfies (2.7) because $R(i, g(i))$ is the one-stage return and $\alpha \sum_j P_{ij}(g(i))V_g(j)$ is the expected additional return obtained by conditioning on the next state visited. That it is the unique solution follows exactly as in the proof of Proposition 2.3. (In fact, we can use Proposition 2.3 directly by considering a problem in which $g(i)$ is the only action available in state i, so V_g is the optimal value function for that problem.)

Remark: In operator notation, (2.7) states that

$$T_g V_g = V_g.$$

3. Method of Successive Approximations

It follows from Theorem 2.2 that, if we could determine the optimal value function V, then we would know the optimal policy—it would be the stationary policy that, when in state i, chooses an action that maximizes

$$R(i, a) + \alpha \sum_j P_{ij}(a)V(j).$$

In this section we show how V can be obtained as a limit of the *n*-stage optimal return.

As a prelude, note that for any policy π and initial state j,

$$\left| E_\pi[\text{return from time } (n + 1) \text{ onwards} | X_0 = j] \right|$$

$$= \left| E_\pi \left[\sum_{i=n+1}^{\infty} \alpha^i R(X_i, a_i) | X_0 = j \right] \right|$$

$$\leq \frac{\alpha^{n+1} B}{1 - \alpha}. \tag{3.1}$$

The method of successive approximations is as follows: Let $V_0(i)$ be any arbitrary bounded function, and define V_1 by

$$V_1(i) = \max_a \left[R(i, a) + \alpha \sum_j P_{ij}(a)V_0(j) \right].$$

In general, for $n > 1$, let

$$V_n(i) = \max_a \left[R(i, a) + \alpha \sum_j P_{ij}(a)V_{n-1}(j) \right].$$

It is worthwhile noting that V_n is the maximal expected discounted return of an n-stage problem that confers a terminal reward $V_0(j)$ if the process ends in state j. The following proposition shows that V_n converges uniformly to V as $n \to \infty$.

Proposition 3.1

(i) If $V_0 \equiv 0$, then $|V(i) - V_n(i)| \le \alpha^{n+1}B/(1-\alpha)$.

(ii) For any bounded V_0, $V_n(i) \to V(i)$ uniformly in i as $n \to \infty$.

Proof: Suppose $V_0 \equiv 0$, so $V_n(i)$ equals the maximal expected return in an n-stage problem starting in i. Now, for the α-optimal policy f,

$$V(i) = E_f(\text{return during first } n\text{-stages}) + E_f(\text{additional returns})$$

$$\le V_n(i) + \alpha^{n+1}B/(1-\alpha),$$

where the inequality follows from (3.1) and the definition of V_n. To go the other way, note that V must be larger than the expected return of the policy that uses the n-stage optimal policy for the first n-stages and any arbitrary policy for the remaining time. Hence,

$$V(i) \ge V_n(i) + E[\text{additional return from } (n+1) \text{ onwards}]$$

$$\ge V_n(i) - \alpha^{n+1}B/(1-\alpha),$$

which, together with the preceding, proves (i).

To prove (ii) let V_n^0 denote V_n when $V_0 \equiv 0$. Then for any bounded V_0 we leave it for the reader to show that

$$|V_n(i) - V_n^0(i)| \le \alpha^n \sup_j |V_0(j)|,$$

which together with (i) proves the result. $\square$

EXAMPLE 3.1 *A Machine Replacement Model*   Suppose that at the beginning of each time period a machine is inspected and its condition or state is noted. After observing this state, a decision as to whether or not to replace the machine must be made. If the decision is to replace, then a cost R is immediately incurred and the state at the beginning of the next time period is 0, the state of a new machine. If the present state is i and a decision not to replace is made, then the state at the beginning of the next time period will be j with probability P_{ij}. In addition, each time the machine is in state i at the beginning of a time period, an operating cost $C(i)$ is incurred.

Let $V(i)$ denote the minimal expected total α-discounted *cost*, given that the initial state is i. Then V satisfies the optimality equation

$$V(i) = C(i) + \min\left[R + \alpha V(0), \alpha \sum_j P_{ij}V(j)\right]. \tag{3.2}$$

Under what conditions on $C(i)$ and the transition probability matrix $\underline{P} = [P_{ij}]$ will $V(i)$ be increasing in i? First, it is clear that the operating costs must be increasing, so let us assume the following condition.

Condition 1: $C(i)$ is increasing in i.

However, Condition 1 by itself is insufficient to imply that $V(i)$ is increasing in i. It is possible for states i and $\bar{\imath}$ (where $i \geq \bar{\imath}$) that whereas i has a higher operating cost than $\bar{\imath}$ it might take the process into a better state than would $\bar{\imath}$. To ensure that this does not occur, we suppose that, under no replacement, the next state from i is stochastically increasing in i. That is, we have the following condition.

Condition 2: For each k, $\sum_{j=k}^{\infty} P_{ij}$ increases in i.

In other words, if T_i is a random variable representing the next state visited after i (assuming no replacement) then $P(T_i = j) = P_{ij}$, so Condition 2 states that[†]

$$T_i \underset{\text{st}}{\leq} T_{i+1}, \qquad i = 0, 1, \ldots.$$

Because this is equivalent to $E(f(T_i))$ increasing in i for all increasing functions f, Condition 2 is equivalent to the statement that

$$\sum_j P_{ij}f(j) \text{ increases in } i, \qquad \text{for all increasing } f.$$

We now prove by induction that, under Conditions 1 and 2, $V(i)$ increases in i.

Proposition 3.2 Under Conditions 1 and 2, $V(i)$ increases in i.

Proof: Let

$$V_1(i) = C(i),$$

and for $n > 1$

$$V_n(i) = C(i) + \min\left[R + \alpha V_{n-1}(0), \alpha \sum_j P_{ij}V_{n-1}(j)\right].$$

---

† See the Appendix for a discussion on stochastic order relations.

It follows from Condition 1 that $V_1(i)$ increases in i. Hence, assume that $V_{n-1}(j)$ increases in j, so, from Condition 2, $\sum_j P_{ij}V_{n-1}(j)$ increases in i, and thus $V_n(i)$ increases in i. Hence, by induction, $V_n(i)$ increases in i for all n, and because

$$V(i) = \lim_n V_n(i),$$

the result follows.   □

The structure of the optimal policy is a simple consequence of Proposition 3.2.

Proposition 3.3   Under Conditions 1 and 2, there exists an $\bar{\imath}, \bar{\imath} \leq \infty$, such that the α-optimal policy replaces when the state is i if $i \geq \bar{\imath}$ and does not replace if $i < \bar{\imath}$.

Proof: It follows from the optimality equation (3.2) that it is optimal to replace in i if

$$\alpha \sum_j P_{ij}V(j) \geq R + \alpha V(0).$$

Because $V(j)$ increasing in j implies that $\sum_j P_{ij}V(j)$ increases in i, the result follows, with $\bar{\imath}$ being given by

$$\bar{\imath} = \min\left[i : \alpha \sum_j P_{ij}V(j) \geq R + \alpha V(0)\right],$$

where $\bar{\imath}$ is taken to be ∞ if the preceding set is empty.

4. Policy Improvement

We have seen that once V is determined the optimal policy is the one that, when in state i, chooses the action a to maximize $R(i, a) + \alpha \sum_j P_{ij}(a)V(j)$. Suppose that for some stationary policy g we have computed V_g, the expected return under g; and suppose that we now define h to be the policy that, when in state i, selects the action that maximizes $R(i, a) + \alpha \sum_j P_{ij}(a)V_g(j)$. How good is h compared with g? We now show that h is at least as good as g, and if it is not strictly better than g for at least one initial state, then g and h are both optimal.

Proposition 4.1 Let g be a stationary policy with expected return V_g and let h be the policy such that

$$R(i, h(i)) + \alpha \sum_j P_{ij}(h(i))V_g(j) = \max_a \left[R(i, a) + \alpha \sum_j P_{ij}(a)V_g(j) \right]. \quad (4.1)$$

Then

$$V_h(i) \geq V_g(i) \qquad \text{for all } i,$$

and if $V_h(i) = V_g(i)$ for all i, then $V_g = V_h = V$.

Proof: Because

$$\max_a \left[R(i, a) + \alpha \sum_j P_{ij}(a)V_g(j) \right]$$

$$\geq R(i, g(i)) + \alpha \sum_j P_{ij}(g(i))V_g(j) = V_g(i),$$

it follows from (4.1) that

$$R(i, h(i)) + \alpha \sum_j P_{ij}(h(i))V_g(j) \geq V_g(i) \qquad \text{for all } i. \quad (4.2)$$

This inequality states that using h for one stage and then switching to g is better than using g throughout. However, because we can repeat this argument after the first stage (that is, at the moment when the first policy is about to switch to g), we see that using h for two stages and then switching to g is better than using g. Repeating this argument shows that using h for n stages and then switching to g is better than using only g; that is,

$$E_h \left[\sum_{j=0}^{n-1} \alpha^j R(X_j, a_j) \Big| X_0 = i \right] + \alpha^n E_h[V_g(X_n)|X_0 = i] \geq V_g(i).$$

Letting $n \to \infty$ gives

$$V_h(i) \geq V_g(i).$$

Now suppose that $V_h(i) = V_g(i)$ for all i. Then, because

$$R(i, h(i)) + \alpha \sum_j P_{ij}(h(i))V_g(j) = V_h(i),$$

we see from (4.1) (upon substituting V_h for V_g) that

$$V_h(i) = \max_a \left[R(i, a) + \alpha \sum_j P_{ij}(a)V_h(j) \right].$$

Hence, V_h satisfies the optimality equation, and by uniqueness (Proposition 2.3), we conclude that $V_h = V$. □

  Remark: In terms of operator notation, we have from (4.2) that

$$T_h V_g \geq V_g,$$

and successively applying T_h to both sides of the preceding inequality gives

$$T_h^n V_g \geq T_h^{n-1} V_g \geq \cdots \geq V_g,$$

implying (letting $n \to \infty$) that

$$V_h \geq V_g.$$

  The preceding result gives us a computational approach to obtaining the optimal policy when the state space is finite. For instance, let the states be $1, 2, \ldots, n$. The optimal policy can be obtained by first choosing any stationary policy g. We then compute V_g as the unique solution of the set of equations

$$V_g(i) = R(i, g(i)) + \alpha \sum_j P_{ij}(g(i)) V_g(j), \qquad i = 1, \ldots, n.$$

Once we have solved this set of n equations in n unknowns and have thus obtained V_g, we then improve g by defining h as the policy that, when in state i, selects the action that maximizes $R(i, a) + \alpha \sum_j P_{ij}(a) V_g(j)$. We next solve for V_h and then improve h, and so on. Because there are only a finite number of possible stationary policies when the state space is finite, we shall eventually reach one for which no strict improvement is possible. This will be the optimal policy.

5. Solution by Linear Programming

If u is a bounded function on the state space satisfying

$$u(i) \geq \max_a \left[R(i, a) + \alpha \sum_j P_{ij}(a) u(j) \right], \qquad i \geq 0, \qquad (5.1)$$

then we shall show that $u \geq V$.

Proposition 5.1   If u satisfies (5.1), then

$$u(i) \geq V(i) \qquad \text{for} \quad \text{all} \quad i.$$

Proof: Consider the usual model with the additional proviso of a stop action that, if exercised when in state i, earns one a terminal reward $u(i)$ and ends the problem. Now, (5.1) states that, for any initial state, stopping immediately is better than doing anything else for one stage and then stopping. If something else (aside from stopping) is done at the initial stage, then, also by (5.1), it is better to stop after the initial stage than it is to do anything else and then stop. Hence, stopping immediately is better than doing anything else for two stages and then stopping. Repeating this shows that stopping immediately is better than doing anything for n stages and then stopping. That is, for any policy π,

$$u(i) \geq E_\pi[n\text{-stage return}|X_0 = i] + \alpha^n E_\pi[u(X_n)|X_0 = i],$$

and upon letting $n \to \infty$, we obtain

$$u(i) \geq V_\pi(i),$$

which implies the result.   □

Remark: If we define the operator T mapping bounded functions on the state space into itself by $T = \max_g T_g$; that is,

$$(Tu)(i) = \max_a \Big[R(i, a) + \alpha \sum P_{ij}(a)u(j) \Big],$$

then (5.1) states that $u \geq Tu$. Applying T to both sides of this inequality gives $Tu \geq T^2u$, so $u \geq T^2u$. Continuing this gives $u \geq T^nu$, and letting $n \to \infty$ and using Proposition 3.1, which states that $T^nu \to V$, gives the result.

Because the optimal value function V satisfies the inequality (as it also satisfies it with equality), it follows from Proposition 5.1 that it is the smallest function that satisfies (5.1). Hence, letting β be such that $0 < \beta < 1$, it follows that V will be the unique solution of the optimization problem

$$\min_u \left[\sum_{i=0}^{\infty} \beta^i u(i) \right],$$

subject to $u(i) \geq \max_a \Big[R(i, a) + \alpha \sum_j P_{ij}(a)u(j) \Big], \qquad i \geq 0,$

or, equivalently,

$$\min_u \left[\sum_{i=0}^{\infty} \beta^i u(i) \right],$$

subject to $u(i) \geq R(i, a) + \alpha \sum_j P_{ij}(a)u(j),$ $i \geq 0,$ $a \in A.$

However, this is a linear programming problem and, at least in the case of a finite state space, can be solved by a technique known as the simplex algorithm. In fact, in the case of a finite state space we can let $\beta = 1$ because we only imposed the condition $0 < \beta < 1$ to keep the objective function finite.

   Remark: For finite-state problems we thus have two possible computational approaches. The linear programming solution just presented and the policy improvement technique of the previous section.

6. Extension to Unbounded Rewards

To ensure that $V_\pi(i)$, defined by

$$V_\pi(i) = E_\pi \left[\sum_{n=0}^{\infty} R(X_n, a_n)\alpha^n \big| X_0 = i \right],$$

is well defined, we have assumed up to now that $R(i, a)$ is bounded. This can be generalized. For instance, suppose that for each i there exist numbers B_i and a constant k such that, starting in i, the expected reward at time $n - 1$ is bounded by $B_i n^k$, $n \geq 1$. That is, for $n \geq 1$, given that $X_0 = i$,

$$E_\pi \big[|R(X_{n-1}, a_{n-1})| \big] \leq B_i n^k, \qquad \text{for all policies } \pi. \qquad (6.1)$$

Under this condition it follows that, conditional on $X_0 = i$,

$$\left| E_\pi \left[\sum_{n=0}^{\infty} R(X_n, a_n)\alpha^n \right] \right| \leq B_i \sum_{n=0}^{\infty} \alpha^n (n + 1)^k < \infty,$$

so V_π remains well defined.

Letting f denote the policy chosen by the optimality equation, then, as in the proof of Theorem 2.2.,

$$V(i) = E_f(n\text{-stage return}) + \alpha^n E_f(V(X_n)), \qquad (6.2)$$

where these expectations are conditional on $X_0 = i$. Now, by condition (6.1) we have

$$\left|\alpha^n E_f(V(X_n))\right| \leq B_i \alpha^n \sum_{j=0}^{\infty} \alpha^j (n + 1 + j)^k$$

$$= B_i \alpha^n \sum_{j=0}^{\infty} \alpha^j \sum_{l=0}^{k} \binom{k}{l} (n + 1)^l j^{k-l}$$

$$\to 0 \qquad \text{as} \quad n \to \infty,$$

where the limit result follows because, for each $l = 0, 1, \ldots, k$,

$$\alpha^n (n + 1)^l \sum_{j=0}^{\infty} \alpha^j j^{k-l} \to 0 \qquad \text{as} \quad n \to \infty.$$

Hence, we see from (6.2), letting $n \to \infty$, that

$$V(i) = V_f(i),$$

so, as in the bounded case, f is optimal.

Policy improvement remains valid in the sense that the improved policy is at least as good as the original. That is, if g is a stationary policy and h is defined so that

$$R(i, h(i)) + \alpha \sum_j P_{ij}(h(i)) V_g(j) = \max_a \left[R(i, a) + \alpha \sum P_{ij}(a) V_g(j) \right],$$

then, exactly as in Proposition 4.1, we have

$$E_h[n\text{-stage return} | X_0 = i] + \alpha^n E_h[V_g(X_n) | X_0 = i] \geq V_g(i).$$

Letting $n \to \infty$ we obtain, using (6.1),

$$V_h(i) \geq V_g(i).$$

Hence, h is at least as good as g. However, it is no longer true that $V_g = V_h$ implies that h and g are optimal. The reason is that it is not necessarily true that V is the unique solution of the optimality equation (in the bounded reward case we could deal with *bounded* functions, and we showed that in the bounded case V is the unique bounded solution of the optimality equation). Of course, if the B_i were bounded, then V would also be bounded, so this result would remain true.

Remark: The preceding remains true even if we suppose that the value of k in (6.1) depends on i; that is, if

$$E_\pi(R(X_{n-1}, a_{n-1})) \le B_i n^{k(i)}, \qquad n \ge 1,$$

where this expectation is conditional on $X_0 = i$.

Problems

1. Consider a problem in which one is interested in maximizing the total expected return. However, suppose that at the end of each time period there is a probability α that the problem ends. Show that this is equivalent to using an infinite-stage discounted-return criterion.

2. Prove the statements given in the technical remark following Theorem 2.2; that is, show

    $$u \le v \Rightarrow T_g u \le T_g v,$$

    and, for bounded u,

    $$T_g^n u \to V_g.$$

3. Prove for any bounded V_0 that, in the successive approximation scheme,

    $$\left| V_n(i) - V_n^0(i) \right| \le \alpha^n \sup_j \left| V_0(j) \right|.$$

4. A quality control model: Consider a machine that can be in one of two states; good or bad. Suppose that the machine produces an item at the beginning of each day. The item produced is either good (if the machine is good) or bad (if the machine is bad). Suppose that once the machine is in the bad state, it remains in that state until it is replaced. However, if it is in the good state at the beginning of a day, then with probability γ it will be in the bad state at the beginning of the next day.

    We further suppose that after the item is produced we have the option of inspecting the item or not. If the item is inspected and found to be in the bad state, then the machine is instantaneously replaced with a good machine at an additional cost R. Also, the cost of inspecting an item will be denoted by I, and the cost of producing a bad item by C.

Suppose that the process is in state P at time t if P is the posterior probability at t that the machine in use is in the bad state. If the objective is to minimize the total expected α-discounted cost, set this up as a Markov decision problem and write the optimality equation.

5. Show that $V(P)$, the optimal value function for Problem 4, is an *increasing, concave function of P*.

6. Consider a machine that can be in either of two states, good or bad. At the beginning of each day, the machine produces items that are either defective or nondefective. The probability of a defective item is P_1 when in the good state and P_2 when in the bad state. Once in the bad state, the machine remains in this state until it is replaced. However, if the machine is in the good state at the beginning of one day, then with probability γ it will be in the bad state at the beginning of the next day. A decision as to whether or not to replace the machine must be made each day after observing the item produced. Let R be the cost of replacing the machine and let C be the cost incurred whenever a defective item is produced. Set this problem up as a Markov decision model and determine the functional equation satisfied by V. Assume that at time zero there is a known probability that the machine is in the bad state.

7. Prove that for Problem 6 there is a P^* such that the α-optimal policy replaces whenever the present probability that the process is in the bad state is greater than or equal to P^*.

8. We have two coins, a red one and a green one. When flipped, one lands heads with probability P_1 and the other with probability P_2. However, we do not know which of the coins is the P_1 coin. Suppose that initially we believe that the red coin is the P_1 coin with probability p_0. Suppose we receive one unit for each head that appears, and our objective to to maximize our total expected discounted return.
    (a) Determine the optimality equation.
    (b) If $P_1 > P_2$, guess at the optimal policy.

9. Consider the following inventory problem. At the beginning of each day the amount of goods on hand is noted and a decision is

made as to how much to order. The cost for ordering j additional units is $C(j)$, where

$$C(j) = \begin{cases} K + cj & \text{if } j > 0, \\ 0 & \text{if } j = 0. \end{cases}$$

The order is assumed to be immediately filled. After the order has been filled, the daily demand for the product occurs. The demand will be j with probability P_j, $j \geq 0$. If the demand exceeds the present supply, then a penalty cost of A per unit of unmet demand is incurred. It is also assumed that, if the demand exceeds the supply, then the additional demand is backlogged and is filled when additional inventory becomes available (this can be represented as negative inventory). In addition there is an inventory holding cost of h for each item of remaining inventory at the end of a period.

The objective is to minimize the total expected discounted cost over an infinite time horizon when α is the discount factor.

(a)   Set this up as a Markov decision process and write the optimality equation.

Consider now a single-period version of the preceding problem. Let

$$L(j) = A \sum_{k=j}^{\infty} (k - j)P_k + h \sum_{k=0}^{j} (j - k)P_k$$

denote the expected penalty and holding costs if we order to bring inventory up to j.

(b)   Show that $L(j)$ is convex. That is, $L(j + 1) - L(j)$ is nondecreasing in j.

(c)   Show that the optimal policy when the initial inventory is i is to order

$$S - i \quad \text{if } i < s,$$

$$0 \quad \text{if } i \geq s,$$

where S is the value that minimizes $cj + L(j)$ and s is such that $cs + L(s) = K + cS + L(S)$.

10.   Assume that an individual has an initial capital of S_0 units. At the beginning of time period n, $n \geq 0$, the individual has S_n units that must be allocated: consuming C_n units, investing I_n units at a sure rate that will return rI_n by the beginning of the next period, and investing J_n units in a risky venture that will return $Z_n J_n$

units by the next period, where Z_n is a random variable having distribution F. Of course, $C_n + I_n + J_n = S_n$. The utility is in consumption, and consuming c leads to a utility $u(c)$. The objective is to maximize the expectation of $\sum_{n=0}^{\infty} \alpha^n u(C_n)$.

(a) Set this up as a dynamic programming problem. Give the optimality equation.

(b) If $u(c) = c^{\beta}$, $0 < \beta < 1$, show that the optimal policy allocates a fixed proportion of one's current fortune to the three alternatives. Show also, in this case, that $V(s) = Ku(s)$ for some constant K.

(c) Suppose that u is a concave function. If $E[Z] < r$, do you think that no money would be allocated to the risky venture? If so, prove it.

11. Consider a problem with states 0, 1, and 2 possible actions having rewards

$$R(0, 1) = 1, \qquad R(1, 1) = 0,$$
$$R(0, 2) = 2, \qquad R(1, 2) = 0,$$

and transition probabilities

$$\begin{bmatrix} P_{00}(1) & P_{00}(2) \\ P_{10}(1) & P_{10}(2) \end{bmatrix} = \begin{bmatrix} \frac{1}{2} & \frac{1}{4} \\ \frac{2}{3} & \frac{1}{3} \end{bmatrix}.$$

Let $\alpha = \frac{1}{2}$. Starting with $V_0 \equiv 0$, use successive approximations to approximate V by V_3. Then show that the policy obtained by maximizing $R(i, a) + \alpha \sum_j P_{ij}(a) V_3(j)$ is the optimal policy.

Now let f be the policy that chooses action 1 in both states 0 and 1. Show that this improvement of policy f is the optimal policy.

12. Let V_f and V_g denote the return functions for stationary policies f and g, respectively. Let h be the policy that chooses actions to maximize $R(i, a) + \alpha \sum_j P_{ij}(a) \max[V_f(j), V_g(j)]$. Show that $V_h(i) \geq \max[V_f(i), V_g(i)]$ for all i.

13. Let f and g be stationary policies with return functions V_f and V_g. Define the policy h by

$$h(i) = \begin{cases} f(i) & \text{if} \quad V_f(i) \geq V_g(i), \\ g(i) & \text{if} \quad V_f(i) < V_g(i). \end{cases}$$

Show that $V_h(i) \geq \max[V_f(i), V_g(i)]$ for all i.

References

1. Bertsekas, D., *Dynamic Programming and Stochastic Control.* Academic Press, New York, 1976.
2. Blackwell, D., "Discounted dynamic programming," *Ann. Math. Statist.* **36,** 226–235, 1965.
3. Derman, C., "On optimal replacement rules when changes of state are Markovian." In *Mathematical Optimization Techniques*, R. Bellman (ed.). University of California Press, Berkeley, California, 1963.
4. Derman, C., *Finite State Markovian Decision Processes.* Academic Press, New York, 1970.
5. Lippman, S., "On dynamic programming with unbounded rewards," *Management Sci.* **21,** 1225–1233, 1975.

Minimizing Costs— Negative Dynamic Programming

1. Introduction and Some Theoretical Results

In this chapter we again assume a countable state space (which, unless otherwise mentioned, will be taken to be the set of nonnegative integers) and a finite action space. However, we now suppose that if action a is taken when in state i, then an expected nonnegative cost $C(i, a)$, $C(i, a) \geq 0$, is incurred. The objective is to minimize the total expected cost incurred. Because this is equivalent to the total expected return for a problem having a reward function $R(i, a)$ $[= -C(i, a)]$ that is nonpositive, we say that we are in the negative case.

For any policy π, let

$$V_\pi(i) = E_\pi\left[\sum_{n=0}^{\infty} C(X_n, a_n) \,\middle|\, X_0 = i\right].$$

Because $C(i, a) \geq 0$, $V_\pi(i)$ is well defined, though possibly infinite. Thus we no longer assume a discount factor, and we no longer require the one-stage costs to be bounded. Also, let

$$V(i) = \inf_\pi V_\pi(i),$$

and call the policy π^* optimal if

$$V_{\pi^*}(i) = V(i), \qquad i \geq 0.$$

Of course, $V(i)$ might be infinite, in which case all policies would be optimal. As a result, this model is only of interest if $V(i) < \infty$ for at least some values of i.

In exactly the same way as we established Theorem 2.1 in Chapter II, we can show the next theorem.

Theorem 1.1   The Optimality Equation
$$V(i) = \min_a \left[C(i, a) + \sum_j P_{ij}(a) V(j) \right].$$

The main theoretical result for negative dynamic programs is that the policy determined by the optimality equation is optimal. That is, we have the following theorem.

Theorem 1.2   Let f be a stationary policy defined by
$$C(i, f(i)) + \sum_j P_{ij}(f(i)) V(j) = \min_a \left[C(i, a) + \sum_j P_{ij}(a) V(j) \right], \qquad i \geq 0.$$

Then
$$V_f(i) = V(i), \qquad i \geq 0.$$

Proof: It follows from the optimality equation and the definition of f that
$$C(i, f(i)) + \sum_j P_{ij}(f(i)) V(j) = V(i). \tag{1.1}$$

In words, if we consider the usual problem with the addition of a stopping option that costs us $V(j)$ if exercised when in state j and that terminates the problem, then (1.1) states that immediate stopping is equivalent to using f for one stage and then stopping. But if we use f for one stage, then at that moment immediate stopping leads to the same expected cost as using f for another stage and then stopping. Hence, using f for two stages and then stopping leads to the same cost as immediate stopping. Repeating this shows that using f for n stages and then stopping is equivalent to immediate stopping. That is,
$$E_f \left[\sum_{t=0}^{n-1} C(X_t, a_t) \Big| X_0 = i \right] + E_f[V(X_n) | X_0 = i] = V(i).$$

Because $V(i) \geq 0$ (all costs are nonnegative), we see that
$$E_f[V(X_n) | X_0 = i] \geq 0,$$

so

$$E_f\left[\sum_{t=0}^{n-1} C(X_t, a_t) \,\big|\, X_0 = i\right] \leq V(i),$$

and upon letting $n \to \infty$,

$$V_f(i) \leq V(i),$$

which proves the result. $\square$

2. Optimal Stopping Problems

In this section we consider models of the following type: When in state i the decision maker can either elect to stop, in which case he receives $R(i)$ and the problem terminates, or to pay $C(i)$ and continue. If the decision is made to continue, then the next state will be j with probability P_{ij}. We suppose that both the continuation costs $C(i)$ and the termination rewards $R(i)$ are nonnegative.

If we say that the process goes to state ∞ when the stopping action is chosen, then this is a two-action decision process, with action 1 representing the stop action and with

$$C(i, 1) = -R(i), \qquad P_{i,j}(2) = P_{ij},$$
$$P_{i,\infty}(1) = 1, \qquad P_{\infty,\infty}(a) = 1,$$
$$C(i, 2) = C(i), \qquad C(\infty, a) = 0.$$

This decision process does not fit the framework of negative dynamic programming because it is not true that all costs are nonnegative. [The terminal reward $R(i)$ is interpreted as a negative cost.] To transform the process into one that does fit the necessary characteristics of negative dynamic programming, let us start by assuming the following two conditions:

(a) $\inf_i C(i) > 0$,

(b) $\sup_i R(i) < \infty$.

That is, we suppose that the continuation costs are bounded away from 0 and the termination rewards are bounded. Now, under these conditions, let $R = \sup_i R(i)$, and consider a related problem in which, when in state i, we may either stop and receive a terminal reward $R(i) - R$ (or, equivalently, pay a terminal cost $R - R(i) \geq 0$) or else pay a continuation cost $C(i)$ and go on to j with probability $P_{ij}, j \geq 0$.

Now, for any policy π, let V_π and $\overline{V}_\pi$ denote the expected cost functions for the original and the related problems, respectively. It is easy to see that for any policy π that stops with probability 1, we have

$$\overline{V}_\pi(i) = V_\pi(i) + R, \qquad i = 0, 1, 2, \ldots.$$

However, these are the only policies we need consider because, by assumption (a), any policy π that does not stop with probability 1 has $V_\pi(i) = \overline{V}_\pi(i) = \infty$. Hence, any policy that is optimal for the original process is optimal for the related process, and vice versa.

However, the related process is a Markov decision process with nonnegative costs. Thus, by the results of the previous section, an optimal policy exists, and the optimal cost function $\overline{V}$ satisfies

$$\overline{V}(i) = \min\left[R - R(i), C(i) + \sum_j P_{ij}\overline{V}(j)\right], \qquad i \geq 0.$$

Also, the policy that chooses the minimizing actions is optimal. Because $V(i) = \overline{V}(i) - R$, we see that

$$V(i) = \min\left[-R(i), C(i) + \sum_j P_{ij}V(j)\right],$$

and the policy thereby determined is optimal.

Now, let

$$V_0(i) = -R(i),$$

and for $n > 0$,

$$V_n(i) = \min\left[-R(i), C(i) + \sum_j P_{ij}V_{n-1}(j)\right], \qquad i \geq 0.$$

Hence, $V_n(i)$ represents the minimal expected total cost incurred if we start in state i and are allowed a maximum of n stages before stopping. From this interpretation we see that

$$V_n(i) \geq V_{n+1}(i) \geq V(i).$$

Hence,

$$\lim_{n \to \infty} V_n(i) \geq V(i),$$

and we say that the process is stable if

$$\lim_{n \to \infty} V_n(i) = V(i).$$

We now show that conditions (a) and (b) ensure stability.

Proposition 2.1 Under conditions (a) and (b), with

$$R = \sup_i R(i) \quad \text{and} \quad C = \inf_i C(i),$$

then

$$V_n(i) - V(i) \le \frac{(R - C)[R - R(i)]}{(n + 1)C}, \quad \text{for all } n \text{ and all } i.$$

Proof: Let f be an optimal policy and let T denote the random time at which f stops. Also, let f_n be the policy that chooses the same actions as f at times $0, 1, \ldots, n - 1$ but that stops at time n (if it has not previously done so). Then, letting X denote the total cost,

$$V(i) = V_f(i) = E_f(X|T \le n)P(T \le n) + E_f(X|T > n)P(T > n),$$

and

$$\tag{2.1}$$

$$V_n(i) \le V_{f_n}(i) = E_f(X|T \le n]P(T \le n) + E_{f_n}(X|T > n)P(T > n),$$

where all expectations are conditional on $X_0 = i$. Therefore,

$$V_n(i) - V(i) \le [E_{f_n}(X|T > n) - E_f(X|T > n)]P(T > n)$$
$$\le (R - C)P(T > n).$$

To obtain a bound on $P(T > n)$, we note by (2.1) that

$$-R(i) \ge V(i) \ge -RP(T \le n) + [-R + (n + 1)C]P(T > n)$$
$$= -R + (n + 1)CP(T > n).$$

Thus

$$P(T > n) \le \frac{R - R(i)}{(n + 1)C},$$

and the result follows. $\square$

To see that not all stopping problems are stable, consider the following example.

EXAMPLE 2.1a *A Nonstable Stopping Problem*   Let the state space be the set of all integers and let

$$C(i) \equiv 0,$$

$$R(i) = i,$$

$$P_{i,i+1} = \tfrac{1}{2} = P_{i,i-1}.$$

Then $V_0(i) = -i$, and it is immediate by induction that $V_n(i) = -i$. However, until the problem is stopped, the state changes according to a symmetric random walk. But such a Markov chain is null recurrent, and thus we know that, with probability 1, any given state N will eventually be reached. Therefore, starting in state i we can guarantee a final reward of N by using the policy that stops when the process enters N. Because this is true for all N, we see that

$$V(i) = -\infty.$$

Hence, this stopping problem (which satisfies neither condition (a) nor condition (b)] is not stable.

Now let

$$B = \left\{ i: -R(i) \le C(i) - \sum_{j=0}^{\infty} P_{ij} R(j) \right\},$$

$$= \left\{ i: R(i) \ge \sum_{j=0}^{\infty} P_{ij} R(j) - C(i) \right\}.$$

Hence, B represents the set of states for which stopping is at least as good as continuing for exactly one more period and then stopping. The policy that stops the first time the process enters a state in B is called the one-stage look-ahead policy. We now prove the useful result that if B is a closed set of states, then, assuming stability, the one-stage look-ahead policy is optimal.

Theorem 2.2   If the process is stable and if $P_{ij} = 0$ for $i \in B$, $j \notin B$, then the optimal policy stops at i if and only if $i \in B$.

Proof: We shall show that $V_n(i) = -R(i)$ for all $i \in B$ and all n. It follows trivially for $n = 0$, so suppose it for $n - 1$. Then, for $i \in B$,

$$V_n(i) = \min\left[-R(i); C(i) + \sum_{j=0}^{\infty} P_{ij} V_{n-1}(j)\right]$$

$$= \min\left[-R(i); C(i) + \sum_{j \in B} P_{ij} V_{n-1}(j)\right] \qquad \text{(because } B \text{ is closed)}$$

$$= \min\left[-R(i); C(i) - \sum_{j \in B} P_{ij} R(j)\right] \quad \text{(by the induction hypothesis)}$$

$$= -R(i).$$

Hence, $V_n(i) = -R(i)$ for all $i \in B$ and all n. By letting $n \to \infty$ and using the stability hypothesis, we obtain

$$V(i) = -R(i), \qquad \text{for} \quad i \in B.$$

Now, for $i \notin B$, the policy that continues for exactly one stage and then stops has an expected cost

$$C(i) - \sum_{j=0}^{\infty} P_{ij} R(j),$$

which is strictly less than $-R(i)$ (because $i \notin B$). Hence,

$$V(i) = -R(i), \qquad i \in B,$$
$$V(i) < R(i), \qquad i \notin B,$$

and the result follows. □

EXAMPLE 2.1b *Selling an Asset*   Suppose that offers come in daily for an asset we own. Each offer, independent of others, will be j with probability $P_j, j \geq 0$. (An offer of zero can be interpreted to mean that no offer arrived that day.) Once the offer is made it must be accepted or rejected, and for each day the item remains unsold we incur a maintenance cost $C, C > 0$. What policy maximizes our expected return, where by return we mean the accepted value minus the total maintenance cost incurred.

With the state at any time being the offer made at that time, it is intuitive that the optimal policy should be of the following form: accept any offer whose value is at least i^*. However, such a set of states is not closed (because offers can decline), thus this problem will not satisfy the conditions of Theorem 2.2. (Check this last statement directly.)

However, let us change the model somewhat by supposing that we are allowed to recall any past offer. In this case the state at any time will be the maximum offer received by that time. Hence, the transition probabilities P_{ij} are given by

$$
P_{ij} = \begin{cases} 0 & \text{if } j < i, \\ \sum_{k=0}^{i} P_k & \text{if } j = i, \\ P_j & \text{if } j > i. \end{cases}
$$

The stopping set for the one-stage look-ahead policy is

$$
\begin{aligned}
B &= \left\{ i : i \ge i \sum_{j=0}^{i} P_j + \sum_{j=i+1}^{\infty} jP_j - C \right\} \\
&= \left\{ i : C \ge \sum_{j=i+1}^{\infty} (j - i)P_j \right\} \\
&= \{ i : C \ge E[(X - i)^+] \},
\end{aligned}
$$

where X is a random variable representing the offer in a given day, that is, $P(X = j) = P_j$, and $x^+ = \max(x, 0)$. Because $(X - i)^+$ decreases in i, we see that in as much as the state cannot decrease (the maximum offer ever received cannot decrease in time), B is a closed set. Hence, assuming stability [which can be shown to hold when $E(X^2) < \infty$) we see that the optimal policy accepts the first offer that is at least i^*, where

$$
i^* = \min \left\{ i : C \ge \sum_{j=i+1}^{\infty} (j - i)P_j \right\}.
$$

Also, because the optimal policy never recalls a past offer, it is also a legitimate policy for the original problem in which such recall is not allowed. Hence, it must be optimal for that problem also. (This follows because it is clear that the maximal return when no recall is allowed cannot be larger than when recall is allowed).

Example 2.1c *The Burglar Problem*   On each attempt, a burglar is successful with probability p. If successful, his loot will be j with probability P_j, $j \ge 0$, and $\sum_j P_j = 1$. If unsuccessful he loses everything (goes to jail) and the problem ends. If the burglar is allowed to retire at any time and keep all the loot he has accumlated up to that time, what is his optimal policy?

This is a stopping problem in which the state at any time is the burglar's total loot at that time. The optimality equation is

$$V(i) = \max\left[i, p \sum_{j=0}^{\infty} V(i + j)P_j\right], \qquad i > 0.$$

The one-stage look-ahead policy is to stop when in B,

$$B = \{i: i \geq p \sum_j (i + j)P_j\}$$

$$= \left\{i: i \geq \frac{p}{1 - p} E(L)\right\},$$

where $E(L) = \sum_j jP_j$, assumed finite, is the mean loot of a successful burglary. Because the state cannot decrease (let the state be ∞ if the burglar is caught), it follows that B is closed, and thus (assuming stability) the one-stage look-ahead policy is optimal.

EXAMPLE 2.1d *Searching for Distinct Types*   In a certain fishing location there are N types of fish; and each time a fish is caught, then, independent of the types of those previously caught, it will be of type i with probability P_i, $i = 1, \ldots, N$. At each stage the decision maker can either pay a cost C and fish during that time period or he can quit. If he fishes, he will catch a fish with probability α; and if he quits, he will receive a terminal reward $R(n)$ if he has caught n distinct types of fish. Let us determine the optimal policy.

The state at any time is the set S of distinct types of fish that have been caught. The one-stage look-ahead policy would stop at S if $S \in B$,

$$B = \{S: R(|S|) \geq \alpha\bar{P}(S)R(|S| + 1) + [1 - \alpha\bar{P}(S)]R(|S|) - C\},$$

where

$$\bar{P}(S) = \sum_{i \notin S} P_i,$$

and $|S|$ denotes the number of elements in S. We can rewrite B as

$$B = \{S: R(|S| + 1) - R(|S|) \leq C/\alpha\bar{P}(S)\}.$$

If we now suppose that $R(n)$ is a concave function, that is, $R(n + 1) - R(n)$ decreases in n, then because S can only become larger [and thus $R(|S| + 1) - R(|S|)$ can only decrease, whereas $C/\alpha\bar{P}(S)$ can only increase] it follows that B is closed. Hence, the one-stage policy is optimal when R is concave.

3. Bayesian Sequential Analysis

Let $Y_1, Y_2, \ldots$ be a sequence of independent and identically distributed random variables. Suppose we know that the probability density function of the Y_i's is either f_0 or f_1, and we are trying to decide on one of these.

At time t, after observing $Y_1, Y_2, \ldots, Y_t$, we may either stop observing and choose either f_0 or f_1 or we may pay a cost C and observe Y_{t+1}. If we stop observing and make a choice, then we incur a cost 0 if our choice is correct and a cost $L(>0)$ if it is incorrect.

Suppose that we are given an initial probability p_0 that the true density is f_0, and say that the state at t is p if p is the posterior probability at t that f_0 is the true density.

This is easily seen to be a three-action Markov decision process with nonnegative costs and an uncountable state space (namely, $[0, 1]$).

If the state is p and we stop and choose f_0, then our expected cost is $(1 - p)L$, and if we stop and choose f_1, then it is pL. If we take another observation when in state p, then the value observed will be x with probability (density)

$$pf_0(x) + (1 - p)f_1(x),$$

and if the value observed is x, then the next state is

$$X_{t+1} = \frac{pf_0(x)}{pf_0(x) + (1 - p)f_1(x)}.$$

Hence, the optimal cost function $V(p)$ satisfies

$$V(p) = \min\left\{(1 - p)L, pL, C + \int V\left[\frac{pf_0(x)}{pf_0(x) + (1 - p)f_1(x)}\right] \times \left[pf_0(x) + (1 - p)f_1(x)\right] dx\right\}.$$

The following lemma will enable us to determine the structure of the optimal policy.

Lemma 3.1 $V(p)$ is a concave function of p.

Proof: Let $p = \lambda p_1 + (1 - \lambda)p_2$, where $0 < \lambda < 1$, and suppose that the density is originally chosen as follows: A coin having probability λ of landing heads is flipped. If heads appears, then f_0 is chosen as the density with probability p_1, and if tails appears, then it is chosen with probability p_2. Now the best that we can do if we are not to be told the outcome of the coin flip is $V[\lambda p_1 + (1 - \lambda)p_2] = V(p)$. On the other hand, if we are to be told the outcome of the flip, then our minimal expected cost is $\lambda V(p_1) + (1 - \lambda)V(p_2)$. Because this must be at least as good as the case in which we are to be given no information about the coin flip (one possible strategy is to ignore this information apriori), we see that

$$V(p) \geq \lambda V(p_1) + (1 - \lambda)V(p_2),$$

which shows that V is concave.   □

Theorem 3.2   There exists numbers p^* and p^{**}, where $p^* \leq p^{**}$, such that when the state is p, the optimal policy stops and chooses f_0 if $p > p^{**}$, stops and chooses f_1 if $p < p^*$, and continues otherwise (see Fig. 3.1).

Proof: Suppose p_1 and p_2 are such that

$$V(p_i) = (1 - p_i)L, \qquad i = 1, 2.$$

Then, for any $p = \lambda p_1 + (1 - \lambda)p_2$, where $\lambda \in [0, 1]$, we have by Lemma 3.1 that

$$V(p) \geq \lambda V(p_1) + (1 - \lambda)V(p_2) = (1 - p)L.$$

However,

$$V(p) \leq (1 - p)L,$$

and thus

$$V(p) = (1 - p)L.$$

Hence, $\{p: V(p) = (1 - p)L\}$ is an interval. Also, it contains the point $p = 1$ because $(1 - p)L = 0$ at $p = 1$. From this, it follows that it is optimal to stop and choose f_0 whenever the state p is larger than some p^{**}. Similar comments hold for $\{p: V(p) = pL\}$ and the theorem follows.   □

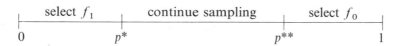

Fig. 3.1. Structure of optimal policy.

4. Computational Approaches

4.1. Successive Approximations

Let

$$V_0(i) = 0,$$

and, for $n > 0$,

$$V_n(i) = \min_a \left[C(i, a) + \sum_j P_{ij}(a) V_{n-1}(j) \right], \qquad i \geq 0.$$

Hence, $V_n(i)$ represents the minimal expected cost for an n-stage problem starting in i. Because all costs are nonnegative, we see that

$$V_n(i) \leq V_{n+1}(i).$$

Let

$$V_\infty(i) = \lim_{n \to \infty} V_n(i).$$

Because $V_n(i) \leq V(i)$ for all n, we see that

$$V_\infty(i) \leq V(i).$$

To prove that $V_\infty = V$ and thus that the optimal value function may be obtained as the limit of the optimal value for the n-stage problem, we first have to show that V is the smallest nonnegative solution of the optimality equation.

Proposition 4.1 If the nonnegative function $u(i)$ is such that

$$u(i) = \min_a \left[C(i, a) + \sum_j P_{ij}(a) u(j) \right], \qquad (4.1)$$

then

$$u(i) \geq V(i).$$

Proof: Let g be the policy determined by the optimality equation (4.1), that is,

$$C(i, g(i)) + \sum_j P_{ij}(g(i)) u(j) = \min_a \left[C(i, a) + \sum_j P_{ij}(a) u(j) \right] = u(i). \quad (4.2)$$

Now consider the usual problem with the addition of a stop action that costs us $u(i)$ if exercised in state i. Then (4.2) states that immediate stopping is equivalent to using g for one stage and then stopping. Repeating this shows that immediate stopping is equivalent to using g for n stages and then stopping. Hence,

$$E_g[n\text{-stage cost}|X_0 = i] + E_g[u(X_n)|X_0 = i] = u(i).$$

Because $u \geq 0$, this yields

$$E_g[n\text{-stage cost}|X_0 = i] \leq u(i),$$

and, letting $n \to \infty$,

$$V_g(i) \leq u(i).$$

Because $V(i) \leq V_g(i)$, the result follows. $\square$

We are now ready to prove the validity of successive approximations.

Theorem 4.2

$$V_\infty(i) = V(i), \qquad i \geq 0.$$

Proof: We have already shown that $V_\infty \leq V$. To go the other way we must show that V_∞ satisfies the optimality equation and thus, by Proposition 4.1, $V \leq V_\infty$. Now,

$$V_\infty(i) = \lim_{n \to \infty} V_n(i)$$

$$= \lim_{n \to \infty} \min_a \left[C(i, a) + \sum_j P_{ij}(a)V_{n-1}(j) \right]$$

$$= \min_a \left\{ \lim_{n \to \infty} \left[C(i, a) + \sum_j P_{ij}(a)V_{n-1}(j) \right] \right\}$$

$$= \min_a \left[C(i, a) + \sum_j P_{ij}(a) \lim_n V_{n-1}(j) \right]$$

$$= \min_a \left[C(i, a) + \sum_j P_{ij}(a)V_\infty(j) \right],$$

where the interchange of limit and minimization is justified because the minimization is over a finite set (the set of actions). The interchange of limit and summation is justified by the Lebesgue monotone convergence theorem [because $V_n(i)$ increases in n]. Hence, V_∞ satisfies the optimality equation, and the proof is complete. $\square$

4.2. Policy Improvement

Let g denote a stationary policy with expected cost function V_g. If we now define the policy h by

$$C(i, h(i)) + \sum_j P_{ij}(h(i))V_g(j) = \min_a \left[C(i, a) + \sum_j P_{ij}(a)V_g(j) \right], \quad (4.3)$$

then h is at least as good as g. That is, we have

Proposition 4.3   A policy h satisfying (4.3) is such that

$$V_h(i) \le V_g(i), \qquad i \ge 0.$$

Proof: It follows from (4.3) that

$$C(i, h(i)) + \sum_j P_{ij}(h(i))V_g(j) \le C(i, g(i)) + \sum_j P_{ij}(g(i))V_g(j) = V_g(i).$$

Hence, the expected cost of using h for one stage and then switching to g is no greater than that of using g. Repeating this shows that the expected cost of using h for n stages and then switching to g is no greater than that of using g. That is,

$$E_h(n\text{-stage total cost} | X_0 = i) + E_h[V_g(X_n) | X_0 = i] \le V_g(i).$$

Because $V_g(i) \ge 0$, we see that

$$E_h[n\text{-stage total cost} | X_0 = i] \le V_g(i);$$

and, letting $n \to \infty$,

$$V_h(i) \le V_g(i). \quad \square$$

Remark: It is not true, as in the discounted case, that if $V_h = V_g$, then g is optimal. If $V_h = V_g$, then because

$$V_h(i) = C(i, h(i)) + \sum_j P_{ij}(h(i))V_h(j),$$

we see from (4.3) that

$$V_h(i) = \min_a \left[C(i, a) + \sum_j P_{ij}(a)V_h(j) \right].$$

That is, V_h satisfies the optimality equation. However, because there need not be a unique solution, we cannot conclude that $V_h = V$. (Proposition 4.1 does not help because it yields $V \le V_h$, which we already know.) Thus the policy improvement algorithm may stop at a nonoptimal policy, whose return satisfies the optimality equation. There are, however, sufficient conditions that would then enable us to conclude that such a policy is optimal (see Problem 11).

5. Optimal Search

5.1. The Classical Optimal Search Model

An object is located in one of n possible locations, being in location i with probability P_i^0, $i = 1, \ldots, n$, $\sum_i P_i^0 = 1$. A search of location i costs C_i, $C_i > 0$, and if the object is present the probability that it will be discovered is α_i, $i = 1, \ldots, n$. That is, $1 - \alpha_i$ is the overlook probability for location i. The objective is to discover the object at minimal expected cost.

This is a decision process whose state at any time is the posterior probability vector $\mathbf{P} = (P_1, \ldots, P_n)$, with P_i representing the posterior probability, given all that has occurred, that the object is in box i. Letting $V(\mathbf{P})$ denote the minimal expected cost function, then the optimality equation is given by

$$V(\mathbf{P}) = \min_i [C_i + (1 - \alpha_i P_i)V(T_i(\mathbf{P}))],$$

where $T_i(\mathbf{P}) = [(T_i(\mathbf{P}))_1, \ldots, (T_i(\mathbf{P}))_n]$ is the vector of posterior probabilities given the prior probability vector $\mathbf{P}$ and given that a search of location i was unsuccessful. Hence,

$$(T_i(\mathbf{P}))_j = P(\text{in } j \,|\, \text{search of } i \text{ unsuccessful})$$

$$= \begin{cases} \dfrac{P_j}{1 - \alpha_i P_i}, & j \neq i, \\[2mm] \dfrac{P_i(1 - \alpha_i)}{1 - \alpha_i P_i}, & j = i. \end{cases}$$

For a given state $\mathbf{P}$, a policy can be regarded as a sequence of locations $(i_1, i_2, \ldots)$ with the interpretation that the locations are searched in that order until the object is found.

For any policy $\delta = (i_1, i_2, \ldots)$, let $V_\delta(\mathbf{P})$ denote the expected cost function under δ. Also, let (i, j, δ) denote the policy that first searches i, then j, and then follows δ.

Lemma 5.1 For any strategy δ such that $V_\delta(\mathbf{P}) < \infty$,

$$V_{(i,j,\delta)}(\mathbf{P}) \leq V_{(j,i,\delta)}(\mathbf{P}) \Leftrightarrow \alpha_i P_i / C_i \geq \alpha_j P_j / C_j.$$

Proof: Suppose $i \neq j$. Then

$$V_{(i,j,\delta)}(\mathbf{P}) = C_i + (1 - \alpha_i P_i)\left[C_j + \left(1 - \frac{\alpha_j P_j}{1 - \alpha_i P_i}\right) V_\delta(T_j T_i \mathbf{P}) \right],$$

and

$$V_{(j,i,\delta)}(\mathbf{P}) = C_j + (1 - \alpha_j P_j)\left[C_i + \left(1 - \frac{\alpha_i P_i}{1 - \alpha_j P_j}\right) V_\delta(T_i T_j \mathbf{P}) \right].$$

Because $T_j T_i \mathbf{P} = T_i T_j \mathbf{P}$, the result follows. $\square$

Proposition 5.2 When in state $\mathbf{P}$, the optimal strategy is to search a location having the maximal value of $\alpha_i P_i / C_i$, $i = 1, \ldots, n$.

Proof: Suppose that

$$\alpha_1 P_1 / C_1 = \max_i \alpha_i P_i / C_i.$$

Now, any policy that never calls for a search of location 1 will have an infinite expected cost (why?), so we can rule out such policies. Next consider any policy δ that does not immediately search location 1 and that searches it for the first time at time j. That is,

$$\delta = (i_1, \ldots, i_{j-1}, 1, i_{j+1}, \ldots).$$

Then, using Lemma 5.1, it is an easy matter (and we leave the details as an exercise) to show that a better policy is obtained by interchanging 1 and i_{j-1}, that is, $\delta' = (i_1, \ldots i_{j-2}, 1, i_{j-1}, \ldots)$ is better than δ. By repeating this argument we see that $(1, i, \ldots, i_{j-1}, i_{j+1}, \ldots)$ is better than δ. Hence, for any policy not initially searching 1, we can find a better one that does start with a search of 1. The result follows. $\square$

5.2. Optimal Search or Stop

Consider the classical model, but, to motivate the search, let us now suppose that a reward R_i, $i = 1, \ldots, n$, is earned if the object is found in the ith box. Suppose also that the searcher may decide to stop searching at any time (for example, he may feel that the rewards are not large enough to justify the searching costs). If the searcher decides to stop before finding the object, then from that point on he incurs no further costs and, of course, receives no reward.

A strategy is any sequence (or partial sequence) $\delta = (\delta_1, \ldots, \delta_s)$, where $\delta_i \in \{1, 2, \ldots, n\}$ for $i = 1, \ldots, s$ and $s \in \{0, 1, 2, \ldots, \infty\}$. The policy δ instructs the searcher to search box δ_i at the ith period and to stop searching if the object has not been found after the sth search (where $s = 0$ means that the searcher stops immediately, and $s = \infty$ means that he does not stop until he finds the object).

If we let $V(\mathbf{P})$ denote the optimal value function when $\mathbf{P} = (P_1, \ldots, P_n)$ is the vector of probabilities as to where the object is, then the optimality equation is given by

$$V(\mathbf{P}) \equiv \min\{0, \min_i [C_i - \alpha_i P_i R_i + (1 - \alpha_i P_i)V(T_i(\mathbf{P}))]\},$$

where $T_i(\mathbf{P})$ is, as before, the posterior probability vector given that a search of i does not find the object. (It should be noted that the preceding is technically not a negative dynamic programming model because of the positive rewards when the object is found. However, it can be transformed into one exactly as we did for the optimal stopping problem of Section 2.)

Let Λ denote the class of strategies $\delta = (\delta_1, \ldots, \delta_s)$ for which $s = \infty$. Any policy $\delta \in \Lambda$ that finds the object with probability 1 will have

$$V_\delta(\mathbf{P}) = E_\delta(L) - \sum_i P_i R_i,$$

where L is the searching cost incurred; any $\delta \in \Lambda$ that has positive probability of never finding the object has $V_\delta(\mathbf{P}) = \infty$. Thus, among the class of policies that never stop searching until the object is found, the one with minimal expected searching cost is best. Hence, by the classical result, the strategy that, when in $\mathbf{P}$, searches the location having maximal value of $\alpha_i P_i / C_i$—call it δ_∞—is optimal among all policies in Λ.

Lemma 5.3   If $\alpha_i P_i R_i > C_i$ for some i, then no optimal strategy stops searching at $\mathbf{P} = (P_1, \ldots, P_n)$. If $\alpha_i P_i R_i \geq C_i$ for some i, then there is an optimal policy that does not stop at $\mathbf{P}$.

Proof: From the optimality equation we have

$$V(\mathbf{P}) \leq C_i - \alpha_i P_i R_i + (1 - \alpha_i P_i) V(T_i(\mathbf{P}))$$
$$< 0 + (1 - \alpha_i P_i) V(T_i(\mathbf{P}))$$
$$\leq 0.$$

Hence, $V(\mathbf{P}) < 0$, and thus no optimal policy stops at $\mathbf{P}$. If $\alpha_i P_i R_i \geq C_i$, then

$$V_i(\mathbf{P}) \equiv C_i - \alpha_i P_i R_i + (1 - \alpha_i P_i) V(T_i(\mathbf{P})) \leq 0.$$

Now, if $V(\mathbf{P}) = 0$ then $V(\mathbf{P}) = V_i(\mathbf{P})$, and so searching i is optimal; whereas if $V(\mathbf{P}) < 0$, then stopping is not optimal. $\square$

Now, for any policy δ, let us denote by (i, j, δ) the policy that first searches i, then (if unsuccessful) j, and then (if unsuccessful) follows δ. In exactly the same way that Lemma 5.1 was proven we can show the following lemma.

Lemma 5.4   For any strategy δ such that $V_\delta(\mathbf{P}) < \infty$,

$$V_{(i,j,\delta)}(\mathbf{P}) \leq V_{(j,i,\delta)}(P) \Leftrightarrow \frac{\alpha_i P_i}{C_i} \geq \frac{\alpha_j P_j}{C_j}.$$

Notation: For any policy $\delta = (\delta_1, \ldots, \delta_s)$ and $t \leq s$, let

$$\mathbf{P}_{\delta,t} = T_{\delta_t} T_{\delta_{t-1}} \cdots T_{\delta_1} \mathbf{P}.$$

Thus $\mathbf{P}_{\delta,t}$ is just the posterior probability vector, given that δ is employed and the item has not been found after t searches.

We are now ready for our main result.

Theorem 5.5   Let $\mathbf{P} = (P_1, \ldots, P_n)$ be the initial state. If

$$\alpha_i P_i / C_i = \max_j \alpha_j P_j / C_j,$$

then the following hold:

(a)   If $\alpha_i P_i R_i \geq C_i$, then there is an optimal policy having $\delta_1 = i$.
(b)   If there does not exist an optimal policy having $\delta_1 = i$, then no optimal strategy ever searches location i.

Proof:

(a) We first show that there is an optimal strategy δ, having $\delta_k = i$ for some $k \leq s$. Suppose that no optimal strategy ever searched i, then for any optimal strategy δ, $(P_{\delta,t})_i \geq P_i$ for all t; so, by Lemma 5.3, the optimal strategy need not stop. But then δ_∞ is optimal, so there would be an optimal strategy with $\delta_1 = i$. Thus there is an optimal strategy δ^* that searches i. Let k be the first time δ^* searches i. If $k \neq 1$, then because

$$(P_{\delta,k-2})_j = \begin{cases} dP_i, & \text{for } j = i, \\ d_j P_j, & \text{for } j \neq i, \end{cases} \quad \text{where } d_j \leq d,$$

it follows that $\alpha_i(P_{\delta,k-2})_i/C_i = \max_j \alpha_j(P_{\delta,k-2})_j/C_j$, so by Lemma 5.4 there is an optimal policy with $\delta_{k-1} = i$. By repeating the argument we see that there is an optimal policy with $\delta_1 = i$.

(b) We have shown by the preceding argument that if an optimal strategy δ has $\delta_k = i$ for some k, then there is an optimal policy with $\delta_1 = i$. $\square$

5.3. Searching for a Moving Target—An Open Problem

Consider the classical model with $n = 2$, that is, there are only two locations. But suppose now that, after each unsuccessful search, the object changes location in accordance with a Markov chain with transition probabilities P_{ij}, $i, j = 1, 2$. That is, an object in location i will move to location j with probability P_{ij}. If we let $V(p)$ denote the optimal value function when the object is initially in location 1 with probability p, then

$$V(p) = \min[C_1 + (1 - \alpha_1 p)V(T_1 p), C_2 + (1 - \alpha_2(1 - p))V(T_2 p)],$$

where $T_i p$, the posterior probability that the object will be in location 1 at the time of the next search (given that a search of location i does not locate it) is given by

$$T_1 p = \frac{(1 - \alpha_1)p}{1 - \alpha_1 p} P_{11} + \frac{1 - p}{1 - \alpha_1 p} P_{21},$$

$$T_2 p = \frac{p}{1 - \alpha_2(1 - p)} P_{11} + \frac{(1 - \alpha_2)(1 - p)}{1 - \alpha_2(1 - p)} P_{21}.$$

It is, however, much more difficult to determine the optimal policy, or even its structure, than in the models of Sections 5.1 and 5.2. In fact, let us suppose that $C_1 = C_2 = 1$, so the objective is to minimize the expected time until the object is found. At first it seems plausible that the optimal policy in this case is simply to search the location that gives the highest probability of finding the object. However, the following counterexample shows that this need not be the case.

EXAMPLE 5.1a   Suppose that $\alpha_1 = \alpha_2 = 1$ and thus that the object is never overlooked. Also suppose

$$P_{12} = 1, \qquad P_{21} = P_{22} = \tfrac{1}{2}.$$

Now, if $p = .55$, then an immediate search of location 1 will discover the object with probability .55, whereas a search of location 2 discovers it with probability .45. However, an unsuccessful search of 2 leads to a certain discovery at the next stage (why?), whereas an unsuccessful search of 1 results in complete uncertainty as to where it will be at the time of the next search. Hence, searching 2 (which results in an expected cost of 1.55) is better than searching 1 (which would at best result in an expected cost of $1 + .45(1.5) = 1.675$).

Whereas the simple myopic policy of searching the location having the highest probability of leading to immediate success is thus not optimal, it has been conjectured that the optimal policy does have the following simple and intuitive structure: for some p^* search location 1 if and only if $p \geq p^*$. However, this remains an unproven conjecture.

Problems

1. Write out the proof of Theorem 1.1.
2. For the model of Example 2.1a show that $V_n(i) = -i$.
3. Suppose that two observations X_1 and X_2 can be taken from a normal distribution having mean 0 and variance 1. If the sampling is terminated after the value X_1 has been observed, then the return is X_1; whereas if it is terminated after X_1 and X_2 have been observed, then the return is $(X_1 + X_2)/2$. Determine the optimal policy and show that the optimal expected return is $1/\sqrt{8\pi}$.
4. Consider the asset-selling problem of Example 2.1b, in which once an offer is rejected, it is lost, but suppose now that the

distribution of possible offers is not known in advance. Suppose, however, that it is initially known that there are n possible distributions F_1, F_2, ..., F_n with given initial probabilities as to which is the true distribution.

(a) Define an appropriate state space and determine the optimality equation.

(b) Determine the one-stage look-ahead policy.

(c) Either prove that the one-stage policy is optimal or give a counterexample.

5. Redo Problem 4 when a rejected offer is not lost but may be recalled at any future time. Does it still remain true (as when the offer distribution is known) that it is never optimal to recall a past offer? Either prove this or give a counterexample.

6. Consider the burglar problem with the change that his success probability changes over time. Specifically, suppose that on his ith burglary he will be successful with probability P_i, $i \geq 1$.

(a) Show that the structure of the optimal policy is as follows: If the burglar has been successful on his first $(i - 1)$ burglaries, $i \geq 1$, then the next one should be attempted if his total loot is less than a_i, for suitably chosen constants $a_i, i \geq 1$.

(b) Give the one-stage look-ahead policy and determine conditions on the P_i, $i \geq 1$, that ensure its optimality. Under these conditions show that a_i, $i \geq 1$, is a monotone sequence.

7. Suppose we are interested in selling our house, and offers come in according to a Markov chain with transition probabilities P_{ij}, $0 \leq i, j \leq N$. If today's offer is i, then tomorrow's will be j with probability P_{ij}. Our objective is to maximize the expectation of the accepted price minus the accumulated maintenance costs, where a maintenance cost C is incurred for each day the house remains unsold. Suppose that

(a) $\sum_{j=k}^{N} P_{ij}$ is nondecreasing in i for each k and

(b) $\sum_{j=0}^{N} (j - i) P_{ij}$ is nonincreasing in i.

Show that the optimal policy is to accept offer i if $i \geq i^*$ and reject it otherwise for some i^*.

8. Consider Example 2.d, but suppose that now the final payoff, when S is the set of distinct types caught, is $R(P(S))$, where $P(S) = \sum_{i \in S} P_i$. Compute the one-stage look-ahead policy and determine conditions on R under which it is optimal.

9. We start a distance of N parking places from our destination. As we drive along we can see only one parking place at a time. Each parking place is, independently, free with probability p. Determine the policy that minimizes the expected distance we need to walk.

10. Give an example in which V_g satisfies the optimality equation and yet g is not optimal.

11. Assume that $C(i, a) \geq 0$. If the state and action spaces are both finite and if f is a stationary policy such that, for each i,

$$V_f(i) < C(i, a) + \sum_j P_{ij}(a)V_f(j) \qquad \text{for} \quad \text{all } a \neq f(i),$$

then show that $V_f = V$.

12. Fill in the details in the proof of Proposition 5.2.

13. For the search and stop model of Section 5.2,
   (a) show that, if $\sum_{i=1}^n C_i/\alpha_i R_i \leq 1$, then the policy that never stops and that, when in state $\mathbf{P}$, searches the location having maximal $\alpha_i P_i/C_i$, is optimal;
   (b) show that, if $\alpha_i R_i < C_i$ for each $i = 1, \ldots, n$, then $V(\mathbf{P}) \equiv 0$;
   (c) show that, if $R_i \equiv R$, then when at state $\mathbf{P}$ there is an optimal policy that either searches the location i with maximal value of $\alpha_i P_i/C_i$ or else stops.

14. For the search and stop model of Section 5.2, prove or give a counterexample to the following statements:
   (a) If $C_i > R_i$, then an optimal policy will never search location i.
   (b) If an optimal policy searches at $\mathbf{P}$, then it searches a location with maximal $\alpha_1 P_i/C_i$.

15. Consider a gambler who, when his fortune is i, is allowed to bet any positive amount j, $j \leq i$, in which case he will either win or lose j with respective probabilities p and $1 - p$. In addition, the gambler is also given the option to work—if so, he works for c units of time and earns 1 unit. Assuming that each gamble takes 1 unit of time, the problem is to minimize the expected time until the worker–gambler attains a fortune of N. (The worker–gambler must work when his fortune is 0). Show that
   (a) if $p \leq \frac{1}{2}$, then always working and never gambling is optimal,
   (b) if $p > \frac{1}{2}$, then always working is optimal if $c < 2/N(2p - 1)$
   Hint: Use the results of Problem 11.

References

1. DeGroot, M. H., *Optimal Statistical Decisions*. McGraw-Hill, New York, 1970.
2. Derman, C., *Finite State Markovian Decision Processes*. Academic Press, New York, 1970.
3. Dynkin, E. B., *Controlled Markov Processes*. Springer-Verlag, New York, 1979.
4. Ross, S. M., "A problem in optimal search and stop," *Oper. Res.* **17** (6), 984–992, 1969.
5. Stone, L. D., *Theory of Optimal Search*. Academic Press, New York, 1975.
6. Strauch, R., "Negative dynamic programming," *Ann. Math. Statist.* **37,** 871–890, 1966.

IV

Maximizing Rewards—
Positive Dynamic Programming

1. Introduction and Main
Theoretical Results

In this chapter we consider models in which one is interested in maximizing expected returns. In particular, assume a countable state space and a finite action space, and suppose that if action a is taken when in state i, then an expected reward $R(i, a)$, assumed nonnegative, is earned.

For any policy π, let

$$V_\pi(i) = E_\pi\left[\sum_{t=0}^{\infty} R(X_t, a_t)\middle| X_0 = i\right], \qquad i \geq 0.$$

Hence, $V_\pi(i)$ is the expected total return earned under π when $X_0 = i$. Because $R(i, a) \geq 0$, $V_\pi(i)$ is well defined, though it may be infinite. Also, let

$$V(i) = \sup_\pi V_\pi(i).$$

A policy π^* is said to be optimal if

$$V_{\pi^*}(i) = V(i) \qquad \text{for} \quad \text{all } i \geq 0.$$

The following may be proven in exactly the same manner as Theorem 2.1 of Chapter II.

Theorem 1.1   The Optimality Equation

$$V(i) = \max_a \left[R(i, a) + \sum_{j=0}^{\infty} P_{ij}(a)V(j) \right], \qquad i \geq 0. \qquad (1.1)$$

Unfortunately, it does not turn out that the policy determined by functional equation (1.1) is an optimal policy. (Try to mimic the proof of Theorem 1.2 of Chapter III and see what goes wrong.) As a matter of fact, it turns out that an optimal policy need not exist. Consider the following.

EXAMPLE 1.1 *Example for Which No Optimal Policy Exists* Suppose there exist two actions and the transition probabilities are given by

$$P_{00}(1) = P_{00}(2) = 1,$$

$$P_{i,i+1}(1) = 1, \qquad i > 0,$$

$$P_{i0}(2) = 1, \qquad i > 0.$$

The rewards are given by

$$R(0, 1) = R(0, 2) = 0,$$

$$R(i, 1) = 0, \qquad\qquad i > 0,$$

$$R(i, 2) = 1 - 1/i, \qquad i > 0,$$

In other words, when in state i we may either get no immediate reward and go to state $i + 1$ or get an immediate reward of $1 - 1/i$ but then collect no future rewards.

It is easy to see that, whereas $V(i) = 1$, $V_\pi(i) < 1$ for all policies π and all i. Hence, an optimal policy does not exist. In fact, the optimality equation is given by

$$V(i) = \max[V(i + 1), 1 - 1/i];$$

and, because $V(i) \equiv 1$, it follows that the policy determined by the optimality equation is the one that always chooses action 1, and thus has an expected return of 0.

Hence, there need not exist an optimal policy (see Problems 1 and 2 for some results relating to sufficient conditions for the existence of an optimal policy). However, one thing that we can and shall show is that, if the return function for a given policy satisfies the optimality equation (1.1), then this policy is optimal.

Theorem 1.2 Let V_f denote the expected return function for the stationary policy f. If V_f satisfies the optimality equation (1.1), then f is optimal. That is, if

$$V_f(i) = \max_a \left[R(i, a) + \sum_j P_{ij}(a) V_f(j) \right], \qquad i \geq 0,$$

then $V_f(i) = V(i)$ for all i.

Proof: The hypothesis of the theorem states that

$$V_f(i) \geq R(i, a) + \sum_j P_{ij}(a) V_f(j), \qquad \text{for all } a. \qquad (1.2)$$

[Of course, equality is attained when $a = f(i)$.] Because the right side of (1.2) is the expected return if we initially take action a and then continue with f, we can interpret (1.2) as stating that using f is better than doing anything else for one stage and then switching to f. But if we do anything else for one stage, then at that point switching to f would be better than doing anything else for another period and then switching to f. Hence, we see that using f is better than doing anything else for two stages and then switching to f. Repeating this argument shows that using f is better than doing anything else for n-stages and then switching to f. That is, for any policy π,

$$V_f(i) \geq E_\pi \left[\sum_{t=0}^{n-1} R(X_t, a_t) \mid X_0 = i \right] + E_\pi[V_f(X_n)].$$

However, because all rewards $R(i, a)$ are nonnegative, it follows that $V_f(i) \geq 0$, so from the preceding we obtain

$$V_f(i) \geq E_\pi \left[\sum_{t=0}^{n-1} R(X_t, a_t) \mid X_0 = i \right].$$

Letting $n \to \infty$ gives

$$V_f(i) \geq V_\pi(i),$$

which proves the result. $\square$

Remarks:

(a) The preceding proof shows that it is not necessary to require that $R(i, a) \geq 0$ for Theorem 1.2 to be valid. A weaker sufficiency condition would be that $V_f(i) \geq 0$ for all i. In fact, an even weaker sufficiency condition would be that

$$\lim_{n} \inf E_{\pi}\big[V_{f}(X_{n})\big|X_{0} = i\big] \geq 0 \qquad \text{for} \quad \text{all } i \text{ and all } \pi.$$

(b)   It follows from Theorems 1.1 and 1.2 that a stationary policy f is optimal if and only if its expected return function satisfies the optimality equation, that is, if for any initial state using f is better than doing anything else for one stage and then switching to f. This gives us a method of checking whether or not a given policy can be optimal and is particularly useful in cases in which there exists an "obvious" optimal policy. This will be well illustrated in the next section.

(c)   Because all rewards are assumed nonnegative, problems fitting this framework are said to be of the positive dynamic programming type.

2. Applications to Gambling Theory

Consider the following situation. An individual possessing i dollars enters a gambling casino. The gambling casino allows any bet of the following form: If you possess i dollars, then you are allowed to bet any positive integral amount less than or equal to i. Furthermore, if you bet j, then you either

(a)   win j with probability p or
(b)   lose j with probability $1 - p$.

What gambling strategy maximizes the probability that the individual will attain a fortune of N before going broke?

In this section we shall show that if $p \geq \frac{1}{2}$ (that is, if you are playing a favorable game), then the optimal strategy is the timid strategy that always bets 1 until your fortune reaches either N or 0. However, if $p \leq \frac{1}{2}$, then the bold strategy, which always bets your present fortune or that part of it that would enable you to get to N if you won, is optimal.

Does this model fit the framework of Section 1? To show that it does we let the state space be the set $\{0, 1, 2, \ldots, N\}$, where we say that the state is i when our present fortune is i. Now, we define the reward structure by

$$R(i, a) = 0, \qquad i \neq N, \qquad \text{for all } a,$$

$$R(N, a) = 1,$$

$$P_{N0}(a) = P_{00}(a) = 1.$$

In other words, a reward of \$1 is earned if and only if our present fortune ever reaches N, and thus the total expected return is just the probability that our fortune ever reaches N. Therefore this problem does indeed fit the framework of the previous section.

To determine an optimal policy, we first note that, if our present fortune is i, then it would never pay to bet more than $N - i$. That is, when in state i, we can limit our choice of bet to $1, 2, \ldots, \min(i, N - i)$.

We know from Theorem 1.2 that, if $U(i)$, the return from some stationary policy, satisfies

$$U(i) \geq pU(i + k) + (1 - p)U(i - k) \tag{2.1}$$

for all $0 < i < N$ and $k \leq \min(i, N - 1)$, then this policy is optimal.

Define the timid strategy to be that strategy that always bets 1. Under this strategy the game becomes the classic gambler's ruin or random-walk model, and $U(i)$, the probability of reaching N before going broke when you start with $i, i < N$, satisfies

$$U(i) = \begin{cases} \dfrac{1 - (q/p)^i}{1 - (q/p)^N} & p \neq \dfrac{1}{2}, \\[2mm] \dfrac{i}{N} & p = \dfrac{1}{2}, \end{cases}$$

where $q = 1 - p$.

Proposition 2.1 If $p \geq \frac{1}{2}$, the timid strategy maximizes the probability of ever attaining a fortune of N.

Proof: If $p = \frac{1}{2}$, then $U(i) = i/N$ trivially satisfies (2.1). When $p > \frac{1}{2}$ we must show that

$$\frac{1 - (q/p)^i}{1 - (q/p)^N} \geq p\left[\frac{1 - (q/p)^{i+k}}{1 - (q/p)^N}\right] + q\left[\frac{1 - (q/p)^{i-k}}{1 - (q/p)^N}\right],$$

or

$$(q/p)^i \leq p(q/p)^{i+k} + q(q/p)^{i-k},$$

or

$$1 \le p(q/p)^k + q(p/q)^k,$$

or, equivalently, that

$$1 \le p[(q/p)^k + (p/q)^{k-1}].$$

Note that this holds if $k = 1$, and thus the result will be proven if we can show that $(q/p)^k + (p/q)^{k-1}$ is an increasing function of k. This is proven by the following lemma.

Lemma 2.2

$$f(x) \equiv \left(\frac{1-p}{p}\right)^x + \left(\frac{p}{1-p}\right)^{x-1}$$

in increasing in x for $x \ge 1$ when $p > \frac{1}{2}$.

Proof:

$$f'(x) = \left(\frac{1-p}{p}\right)^x \log\left(\frac{1-p}{p}\right) + \left(\frac{p}{1-p}\right)^{x-1} \log\left(\frac{p}{1-p}\right)$$

$$= \left[\left(\frac{p}{1-p}\right)^{x-1} - \left(\frac{1-p}{p}\right)^x\right] \log\left(\frac{p}{1-p}\right)$$

$$\ge 0$$

because $p > \frac{1}{2}$ and $x \ge 1$. This proves the lemma and also completes the proof of Proposition 2.1. ☐

Thus when $p \ge \frac{1}{2}$, that is, when the game is favorable for us, then we should continually place the minimum bet until either reaching our goal or going broke. Before looking at the case in which $p < \frac{1}{2}$, let us now suppose that our objective is not to reach some preassigned goal, but rather to maximize our playing time. We now show that if $p \ge \frac{1}{2}$, then timid play stochastically maximizes our playing time. That is, for each n, the probability that we will be able to play n or more times before going broke is maximized by the timid strategy.

Proposition 2.3   If $p \ge \frac{1}{2}$, then timid play stochastically maximizes our playing time.

Proof: By assuming that a reward of 1 is attained if we are able to play at least n times, we see that this problem also fits the framework of the positive case. Hence, we must show that, starting with i, it is better to play timidly than it is to make an initial bet of k, $k \leq i$, and then play timidly. However, this follows because, by Proposition 2.1, the timid strategy maximizes the probability that we will reach $i + k$ before $i - k$, and it takes *at least* one unit of time. More formally, letting $U_n(i)$ denote the probability that we will be able to play at least n times, given that our initial fortune is i and we play timidly, we obtain, by conditioning on the time T that our fortune reaches either $i - k$ or $i + k$ and the value X that is reached, that

$$U_n(i) = E[U_{n-T}(X)]$$

$$\geq E[U_{n-1}(X)]$$

$$= U_{n-1}(i + k)P(X = i + k) + U_{n-1}(i - k)P(X = i - k)$$

$$\geq pU_{n-1}(i + k) + qU_{n-1}(i - k).$$

The first inequality follows from the facts that $U_n(i)$ is a decreasing function of n and $T \geq 1$, whereas the second inequality follows because $P\{X = i + k\} \geq p$ by Proposition 2.1, and

$$U_{n-1}(i + k) \geq U_{n-1}(i - k). \quad \square$$

It turns out that if $p < \frac{1}{2}$, then the timid strategy does not stochastically maximize playing time. Suppose $p = 0.1$ and that we start with an initial fortune of 2. The probability that we will be able to play at least 5 games if we play timidly is $1 - (0.9)^2 - 2(0.9)^3(0.1) = 0.0442$. On the other hand, if we bet 2 initially and then play timidly, then the probability of playing at least 5 games is 0.1. It is true, however, that timid play maximizes our expected playing time.

Proposition 2.4 If $p < \frac{1}{2}$, then timid play maximizes our expected playing time.

Proof: Let $U(i)$ denote the expected number of bets made before we go broke, given that we start with i and always bet 1. To calculate $U(i)$, let X_j denote the winnings on the jth bet and let T denote the number of bets before going broke. Then, because

$$\sum_{j=1}^{T} X_j \equiv -i,$$

we have by Wald's equation that

$$-i = E(X)E(T)$$

or

$$U(i) = E(T) = \frac{i}{1 - 2p}.$$

Because maximizing our expected playing time falls under the positive case (we receive a reward of 1 each time that we are able to continue playing), the result follows because the inequality

$$U(i) \geq 1 + pU(i + k) + qU(i - k), \qquad 1 \leq k \leq i,$$

is equivalent to

$$\frac{i}{1 - 2p} \geq 1 + p\frac{(i + k)}{1 - 2p} + (1 - p)\frac{(i - k)}{1 - 2p},$$

or

$$0 \geq 1 - 2p + pk - (1 - p)k,$$

or

$$k(1 - 2p) \geq 1 - 2p,$$

which follows because $k \geq 1$. $\square$

Thus when $p < \frac{1}{2}$, the timid strategy is optimal when our objective is to maximize the expected amount of time that we are able to play before going broke. However, we shall now show that if our criterion is to reach a fortune of N, then it is the bold strategy that is optimal. In fact, we will show that this is the case even when one is limited in the number of bets allowed.

Define the bold strategy to be the strategy that, if our present fortune is i,

(a)   bets i if $i \leq N/2$,
(b)   bets $N - i$ if $i \geq N/2$.

Let $U_n(i)$ represent the probability of ever reaching N given that we start with i, are allowed at most n bets, and use the bold strategy. By conditioning on the outcome of the first bet it follows that

$$U_n(i) = \begin{cases} pU_{n-1}(2i) & \text{if } i \leq N/2, \\ p + qU_{n-1}(2i - N) & \text{if } i \geq N/2, \end{cases} \qquad (2.2)$$

with boundary conditions $U_n(0) = 0$, $U_n(N) = 1$, $n \geq 0$, $U_0(i) = 0$, $i < N$.

We are now ready for the next proposition.

Proposition 2.5   When $p \leq \frac{1}{2}$, for each $n > 0$, the bold strategy maximizes the probability of attaining a fortune of N by time n.

Proof: By Theorem 1.2 if suffices to prove that

$$U_{n+1}(r) \geq pU_n(r + s) - qU_n(r - s), \qquad s \leq \min(r, N - r)$$

or, equivalently, that

$$U_{n+1}(r) - pU_n(r + s) - qU_n(r - s) \geq 0, \qquad s \leq \min(r, N - r). \quad (2.3)$$

We shall verify this inequality by induction on n. Because it is immediate for $n = 0$, let us assume that

$$U_n(i) - pU_{n-1}(i + k) - qU_{n-1}(i - k) \geq 0, \qquad k \leq \min(i, N - i). \quad (2.4)$$

To establish (2.3) there are four cases we must consider.

Case 1: $r + s \leq N/2$. In this case we have by (2.2) that

$$U_{n+1}(r) - pU_n(r + s) - qU_n(r - s)$$
$$= p[U_n(2r) - pU_{n-1}(2r + 2s) - qU_{n-1}(2r - 2s)].$$

But this is nonnegative, as is seen by using (2.4) with $i = 2r$ and $k = 2s$.

Case 2: $r - s \geq N/2$. The proof is just as in Case 1, except that the second functional equation of (2.2) is used instead of the first.

Case 3: $r \leq N/2 \leq r + s, s \leq r$. In this case we have from (2.2) that

$$U_{n+1}(r) - pU_n(r + s) - qU_n(r - s)$$
$$= p[U_n(2r) - p - qU_{n-1}(2r + 2s - N) - qU_{n-1}(2r - 2s)].$$

Now, $2r \geq r + s \geq N/2$, so the previous inequality can be continued as

$$= p[p + qU_{n-1}(4r - N) - p - qU_{n-1}(2r + 2s - N)$$
$$\quad - qU_{n-1}(2r - 2s)]$$
$$= q[pU_{n-1}(4r - N) - pU_{n-1}(2r + 2s - N) - pU_{n-1}(2r - 2s)]$$
$$= q[U_n(2r - N/2) - pU_{n-1}(2r + 2s - N) - pU_{n-1}(2r - 2s)],$$

$$(2.5)$$

where the last equality follows from (2.2) because $0 \leq 2r - N/2 \leq N/2$. Now, if $s \geq N/4$, then, because $p \leq q$, we have that (2.5) is at least

$$q[U_n(2r - N/2) - pU_{n-1}(2r + 2s - N) - qU_{n-1}(2r - 2s)].$$

However, this last expression is nonnegative by the induction hypothesis (2.4) with $i = 2r - N/2$ and $k = 2s - N/2$.

On the other hand, if $s < N/4$, then, because $p \leq q$, (2.5) is at least

$$q[U_n(2r - N/2) - qU_{n-1}(2r + 2s - N) - pU_{n-1}(2r - 2s)],$$

which is nonnegative by the induction hypothesis (2.4) with $i = 2r - N/2$ and $k = N/2 - 2s$.

Case 4: $r - s \leq N/2 \leq r$. The proof for this case is similar to that of the preceding one and is left as an exercise.

Corollary 2.6 With unlimited time, bold play maximizes the probability of ever reaching N when $p \leq \frac{1}{2}$.

Proof: If we let $U(r)$ denote the probability of reaching N before 0 if we start with r and use the bold strategy, then because

$$U(r) = \lim_{n \to \infty} U_n(r),$$

it follows from (2.3) that

$$U(r) \geq pU(r + s) + qU(r - s), \qquad s \leq \min(r, N - r).$$

Hence, the return from bold play satisfies the optimality equation and is thus, from Theorem 1.2, optimal. □

Remarks: (i) We have thus shown that in red–black gambling models in which one attempts to reach a fortune of N before going broke, timid play is optimal when $p \geq \frac{1}{2}$ and bold play when $p \leq \frac{1}{2}$. Let us call a strategy "stupid" if for some fortune i it calls for a bet greater than $N - i$, and let us agree to restrict attention to nonstupid strategies. Let us imagine that we have an opponent whose fortune is $N - i$ when ours is i and who wins when we lose and vice versa. Because what is best for our opponent is worse for us and vice versa and because our opponent is playing the same game except that his win probability is $1 - p$, we see that

(a)   when $p \geq \frac{1}{2}$, the bold strategy is the worst strategy;
(b)   when $p \leq \frac{1}{2}$, the timid strategy is the worst strategy.

Interestingly, when $p = \frac{1}{2}$, because the bold is both the best and the worse strategy, it follows that all (nonstupid) strategies have the same return. Because the probability of reaching N before 0 when starting with i and playing timidly is, when $p = \frac{1}{2}$, i/N, we see that for any nonstupid strategy π,

$$V_{\pi}(i) = i/N \qquad \text{when} \quad p = \tfrac{1}{2}.$$

(ii)   In Corollary 2.6 we used the result that

$$U(r) = \lim_{n} U_{n}(r).$$

This follows from the continuity property of the probability operator.

3. Computational Approaches to Obtaining V

3.1. Limit of the Maximal Finite-Stage Returns

Let

$$V_0(i) = 0, \qquad i \geq 0,$$

and, for $n > 0$,

$$V_n(i) = \max_{a}\left[R(i, a) + \sum_{j} P_{ij}(a)V_{n-1}(j)\right], \qquad i \geq 0.$$

Hence, $V_n(i)$ represents the maximal expected return in an n-stage problem starting in i. We then have

Proposition 3.1

$$\lim_{n \to \infty} V_n(i) = V(i), \qquad i \geq 0.$$

Proof: Because $R(i, a) \geq 0$, it follows that $V_n(i)$ increases in n, so $\lim_{n} V_n(i)$ exists. Now, for any policy π

$$E(n\text{-stage return under } \pi \,|\, X_0 = i) \leq V_n(i).$$

Letting $n \to \infty$ now gives

$$V_{\pi}(i) \leq \lim_{n} V_n(i).$$

Because this is true for all policies π, we see that

$$V(i) \le \lim_n V_n(i). \tag{3.1}$$

On the other hand, because all rewards are nonnegative, it follows that the maximal n-stage return is no greater than the supremal infinite-stage return, so

$$V_n(i) \le V(i) \qquad \text{for all } n,$$

which, along with (3.1), yields the result. ☐

Hence, one can attempt to obtain (or approximate) V by considering the finite-stage problem and letting the number of stages become large. A second approach is by using linear programming.

3.2. A Linear Programming Solution for V

Let $u(i)$, $i \ge 0$, be a nonnegative function such that

$$u(i) \ge \max_a \left[R(i, a) + \sum_j P_{ij}(a)u(j) \right]. \tag{3.2}$$

Analogous to Proposition 5.1 of Chapter II, we have the next proposition.

Proposition 3.2   A nonnegative function u satisfying (3.2) is such that

$$u(i) \ge V(i), \qquad i \ge 0.$$

Proof: Consider the standard positive dynamic programming problem with the exception that we are given the option of stopping at any time. If we stop when in state i, then we earn a terminal reward $u(i)$. Now, (3.2) states that immediate stopping is better than doing anything else for one stage and then stopping. Repeating this argument shows that immediate stopping is better than doing anything else for n-stages and then stopping. That is, for any policy π,

$$u(i) \ge E_\pi(n\text{-stage return}|X_0 = i) + E_\pi \left[u(X_n)|X_0 = i \right].$$

Because $u \ge 0$, this yields

$$u(i) \ge E_\pi(n\text{-stage return}|X_0 = i)$$

and, upon letting $n \to \infty$,

$$u(i) \geq V_\pi(i).$$

Because this is true for all π, the result follows. $\quad \square$

Suppose now that the state space is finite and that the problem is such that $V(i) < \infty$ for each state i. Then because V also satisfies (3.2) (in fact with equality), it follows from Proposition 3.2 that V is the smallest solution. Hence, V may be obtained as the solution of the following linear program:

$$\text{minimize} \quad \sum_i u(i),$$

$$\text{subject to} \quad u(i) \geq R(i, a) + \sum_j P_{ij}(a)u(j) \quad \text{for} \quad \text{all } i, a,$$

$$u(i) \geq 0, \quad i \geq 0.$$

Problems

1. Let f be a stationary policy determined by the optimality equation (1.1), that is, $f(i)$ is such that

   $$R(i, f(i)) + \sum_j P_{ij}(f(i))V(j) = \max_a \left[R(i, a) + \sum_j P_{ij}(a)V(j) \right].$$

   Show that a necessary and sufficient condition for f to be optimal is for

   $$E_f[V(X_n)|X_0 = i] \to 0 \quad \text{as} \quad n \to \infty \quad \text{for all} \quad i.$$

2. Suppose state 0 is such that

   $$P_{00}(a) = 1 \quad \text{for} \quad \text{all } a,$$

   $$R(0, a) = 0 \quad \text{for} \quad \text{all } a.$$

   Let f be a policy determined by the optimality equation (1.1). Show, by using the results of Problem 1, that if (a) V is bounded and (b) with probability 1, $X_n \to 0$ as $n \to \infty$ when policy f is employed, then f is an optimal policy.

3. Let T map nonnegative functions on the state space into itself as follows:

   $$(Tu)(i) = \max_a \left[R(i, a) + \sum_j P_{ij}(a)u(j) \right].$$

Show that

(a) $TV = V$.
(b) $u \leq v \Rightarrow Tu \leq Tv$.
(c) $(T^n 0)(i) \to V(i)$ as $n \to \infty$, where 0 is the function identically equal to 0.
(d) If the nonnegative function u satisfies $u \geq Tu$, show that $u \geq V$.

4.   If $V_0(i) = u(i)$ and $V_n(i) = \max_a [R(i, a) + \sum_j P_{ij}(a) V_{n-1}(j)]$ for $n > 1$, where u is an arbitrary nonnegative function, is it true that $V_n(i) \to V(i)$? If $u(i) \equiv 0$, is the convergence uniform in i?

5.   Give an example in which the optimality equation does not possess a unique nonnegative solution.

6.   Consider a process that, when in state i, allows us either to stop and receive a terminal nonnegative reward $R(i)$ or continue and go to the next state according to the transition probabilities P_{ij}. Show that $V(i)$ is the smallest function satisfying

$$V(i) \geq R(i) \qquad \text{for all } i,$$

$$V(i) \geq \sum_j P_{ij} V(j) \qquad \text{for all } i.$$

Consider the symmetric random walk with absorbing barriers 0 and N, that is,

$$P_{i,i-1} = P_{i,i+1} = \tfrac{1}{2}, \qquad 0 < i < N,$$

$$P_{00} = P_{NN} = 1.$$

Argue that $V(i)$ is the smallest concave function greater than or equal to $R(i)$, $0 \leq i \leq N$. Find $V(i)$ when $N = 7$, $R(0) = 0$, $R(1) = 4$, $R(2) = 2$, $R(3) = 1$, $R(4) = 2$, $R(5) = 5$, $R(6) = 7$, $R(7) = 0$.

7.   A bet is said to be a (k, c) bet with fairness parameter α if the outcome of the bet is that you either win k with probability $\alpha c/(k + c)$ or lose c with probability $1 - \alpha c/(k + c)$. If $\alpha \geq 1$, then these are favorable bets (calculate your expected gain), and if $\alpha < 1$, then they are unfavorable bets.

Suppose we enter a gambling casino that allows any (k, c) bet when k and c are positive integers and α is a fixed constant greater than 1. Does the timid strategy, which always bets $(1, 1)$, maximize our probability of attaining a fortune of N before going broke?

8. In Problem 7, show that if $\alpha < 1$, then the timid strategy maximizes the expected playing time until you go broke.

9. Consider a gambling casino allowing any (k, c) bet with fairness parameter $\alpha = 1$. Let $U(i, n)$ represent the probability that you will still have some money left after the nth gamble if you start with i and always bet $(1, 1)$. Show that $U(i, n)$ is a concave function of i for all n.

    Hint: Prove it first for all i of the form $i = \lambda i' + (1 - \lambda)i''$ first for $\lambda = \frac{1}{2}$ and then (by a standard argument) for all $\lambda \in (0, 1)$.

10. Use the results of Problem 9 to show that the timid strategy of always betting $(1, 1)$ maximizes the probability that you will survive n gambles, for all n, when $\alpha = 1$.

11. Do Problems 9 and 10 remain valid when $\alpha > 1$?

12. Suppose that the gambling casino allows you to make any bet such that, when your present fortune is i, the outcome of the bet is a random variable X satisfying

    (a) $X \geq -i$,
    (b) $|X| \geq 1$,
    (c) $E(X) \leq \alpha - 1$.

    Show that if $\alpha < 1$, then the strategy that always bets 1 to either win or lose 1 with respective probabilities $\alpha/2$ and $1 - \alpha/2$ maximizes your expected playing time until you go broke.

13. Let $U(r)$ be the probability of reaching 1 before 0 when one's initial fortune is r and bold play is employed. Show that for $p \leq \frac{1}{2}$,

    $$U(r + s) \geq U(r) + U(s).$$

    This is a weaker condition than convexity, which U does not satisfy. Give a heuristic argument for why U is not convex.

14. For an initial fortune $i < N/2$, show that the policy that uses bold play to reach the goal $N/2$ and if successful then bets the $N/2$ is also optimal for the problem of reaching N before 0.

15. Does policy improvement work in the positive case. That is, for any policy g with expected return function V_g, if we define the policy h by

    $$R(i, h(i)) + \sum_j P_{ij}(h(i))V_g(j) = \max_a \left[R(i, a) + \sum_j P_{ij}(a)V_g(j)\right],$$

is it true that $V_h \geq V_g$? Prove or give a counterexample. If it is not true, give a sufficient condition under which it would be valid.

Notes and References

Our theoretical approach follows that of Blackwell [2]. The early gambling models of Section 2 comes from Ross [6]. The deeper result on the optimality of bold play when $p < 1/2$ is due to Dubins and Savage [4], though the modification of their proof that we have given is due to Dynkin [5]. For a discussion of the probability results used in this chapter, the interested reader should see Ross [7].

1. Bertsekas, D., *Dynamic Programming and Stochastic Control.* Academic Press, New York, 1976.
2. Blackwell, D., "Positive dynamic programming," *Proc. 5th Berkeley Symp. Math. Statist. Probab.* **1,** 415–418, 1967.
3. Derman, C., *Finite State Markovian Decision Processes.* Academic Press, New York, 1970.
4. Dubins, L. and Savage, L. J., "*How to Gamble If You Must.*" McGraw-Hill, New York, 1965.
5. Dynkin, E. B., *Controlled Markov Process.* Springer-Verlag, New York, 1979.
6. Ross, S. M., "Dynamic programming and gambling models," *Adv. Appl. Probab.* **6,** 593–606, 1974.
7. Ross, S. M., *Stochastic Processes.* Wiley, New York, 1982.

V

Average Reward Criterion

1. Introduction and Counterexamples

In this chapter we suppose that a bounded reward $R(i, a)$ is earned whenever action a is chosen in state i and that the objective is to maximize the long-run expected average reward per unit time. The set of possible actions is assumed finite, and for any policy π and initial state i, we let

$$\phi_\pi(i) = \lim_{n \to \infty} \frac{E_\pi\left[\sum_{j=0}^n R(X_j, a_j) \mid X_0 = i\right]}{n + 1},$$

where X_j and a_j are, respectively, the state and action chosen at time j. If this limit does not exist, let us agree to define $\phi_\pi(i)$ by the lim inf. Hence, $\phi_\pi(i)$ represents the average expected return per unit time when π is employed and the initial state is i.

We say that π^* is *average-reward optimal* if

$$\phi_{\pi^*}(i) = \max_\pi \phi_\pi(i) \qquad \text{for} \quad \text{all } i.$$

The first question we shall consider is whether an optimal policy need necessarily exist; and we answer it by our first counterexample.

COUNTEREXAMPLE 1.1 *Optimal Policies Need Not Exist* Let the state space be the set $\{1, \bar{1}, 2, \bar{2}, \ldots, n, \bar{n}, \ldots\}$. There are two actions and the transition probabilities are given by

$$P_{i, i+1}(1) = P_{i, i}(2) = 1, \qquad i \geq 1,$$
$$P_{i, i}(1) = P_{i, i}(2) = 1, \qquad i \geq 1.$$

89

The rewards depend only on the state and are given by

$$R(i, a) = 0, \qquad\qquad i \geq 1,$$

$$R(i, a) = 1 - \frac{1}{i}, \qquad i \geq 1.$$

In words, when in state i, we can either receive nothing and go to state $i + 1$ or we can elect to go to state i and, as a result, earn $1 - 1/i$ for each time period thereafter.

Clearly, for any policy π, $\phi_\pi(1) < 1$. However, because we can obtain a return arbitrarily close to 1 (choosing action 2 for the first time when in state n leads to an average return of $1 - 1/n$), it follows that an optimal policy does not exist.

In this counterexample, although an optimal policy does not exist, it is true that for any policy there is a stationary policy that does at least as well. If this were always to be the case, then it would suffice to restrict attention to the class of stationary policies. (It certainly seems intuitive that the "optimal" action at any time should depend only on the state at that time.) However, we now show that this need not be true.

COUNTEREXAMPLE 1.2 *Stationary Policies Do Not Suffice*   Let the state space be the positive integers and suppose there are two actions. For $i \geq 1$, the rewards and transition probabilities are given by

$$P_{i,i+1}(1) = 1 = P_{i,i}(2),$$

$$R(i, 1) = 0,$$

$$R(i, 2) = 1 - \frac{1}{i}.$$

In words, when in state i, we can either receive nothing and go to $i + 1$ or receive $1 - 1/i$ and stay in i.

Suppose that $X_0 = 1$, and let π be any stationary policy. There are two cases.

Case 1: π always chooses action 1   In this case, $\phi_\pi(1) = 0$.

Case 2: π chooses action 2 for the first time at state n In this case the process will go from state 1 to 2 to 3 to $\cdots$ to n. However, once in n, action 2 is chosen, so the process will never leave that state and $1 - 1/n$ will be earned from there on. Hence,

$$\phi_\pi(1) = 1 - \frac{1}{n} < 1.$$

Hence, for any stationary policy π, $\phi_\pi(1) < 1$.

Now let π^* be the nonstationary policy that, when the process first enters state i, chooses action 2 i consecutive times and then chooses action 1. Because the initial state is $X_0 = 1$, it follows that the successive rewards received will be

$$0, 0, \tfrac{1}{2}, \tfrac{1}{2}, 0, \tfrac{2}{3}, \tfrac{2}{3}, \tfrac{2}{3}, 0, \tfrac{3}{4}, \tfrac{3}{4}, \tfrac{3}{4}, \tfrac{3}{4}, 0, \tfrac{4}{5}, \ldots.$$

The average value of such a sequence is clearly equal to 1, and thus $\phi_{\pi^*}(1) = 1$. Hence, the nonstationary policy π^* is strictly better than every stationary policy.

The preceding counterexample is one in which there is no best stationary policy, and by constantly switching from one stationary policy to a better one, we are able to obtain a nonstationary policy that attains the limiting average return of these stationary policies. This raises the question of whether nonstationary policies may be needed only because they sometimes allow us to obtain the limit of the returns from a sequence of stationary policies. If this were to be true, then for any nonstationary policy we would find a stationary one whose average return is arbitrarily close. However, our next counterexample presents a situation in which no stationary policy has a return within ϵ of a given nonstationary policy.

COUNTEREXAMPLE 1.3 *No Stationary Policy Is Within ϵ of Optimality* Let the states by given by $1, 1', 2, 2', \ldots, n, n', \ldots, \infty$. In state n, $1 \leq n < \infty$, there are two actions, with transition probabilities given by

$$P_{n,n+1}(1) = 1,$$

$$P_{n,n'}(2) = \alpha_n = 1 - P_{n,\infty}(2).$$

In state n', there is a single action having transition probabilities

$$P_{n',(n-1)'} = 1, \qquad n \geq 2,$$

$$P_{1',1} = 1.$$

State ∞ is an absorbing state and once entered it is never left, that is,

$$P_{\infty,\infty} = 1.$$

The rewards depend only on the state and are given by

$$R(n, a) = 0 \quad \text{for all } n = 1, 2, \ldots, \infty \quad \text{and all actions } a,$$
$$R(n', a) = 2 \quad \text{for all } n \geq 1 \quad \text{and all } a.$$

The values α_n are chosen to satisfy

$$\alpha_n < 1, \quad \text{and} \quad \prod_{n=1}^{\infty} \alpha_n = \tfrac{3}{4}.$$

In words, when in state n, we can elect to go to state $n + 1$ or we can elect to try to reach state n'. If the attempt to reach n' is made, it will be successful with probability α_n and will earn us 2 units for each of the next n time periods. However, if unsuccessful, then state ∞ is entered and all future rewards equal 0. In addition, there is no reward for being in any of the unprimed states $1, 2, \ldots$. Suppose the initial state is 1. If a stationary policy is employed, then with probability 1, a reward of zero will be received in all but a finite number of time periods. This follows because, under a stationary policy, each time the process enters state 1 there is a fixed positive probability that the process will never again reenter that state. Therefore, under a stationary policy, the average reward will, with probability 1, equal zero. Hence, the average expected reward will also equal zero.

Now let R be the nonstationary policy that initially chooses action 2 and, on its nth return to state 1, chooses action 1 n times and then chooses action 2. The average reward under this policy will equal

$$0 \quad \text{with probability } 1 - \prod_{n=1}^{\infty} \alpha_n,$$

$$1 \quad \text{with probability } \prod_{n=1}^{\infty} \alpha_n.$$

This is true because $\prod_{n=1}^{\infty} \alpha_n$ represents the probability that, under R, the process will never enter state ∞. Hence, the average expected reward under R is $\tfrac{3}{4}$.

2. Existence of an Optimal Stationary Policy

In this section we shall determine conditions that result in the existence of an optimal policy that is stationary. We begin with a theorem that can be regarded as the average-return version of the optimality equation.

Theorem 2.1   If there exists a bounded function $h(i)$, $i \geq 0$, and a constant g such that

$$g + h(i) = \max_a \left[R(i, a) + \sum_{j=0}^{\infty} P_{ij}(a)h(j) \right], \qquad i \geq 0, \qquad (2.1)$$

then there exists a stationary policy π^* such that

$$g = \phi_{\pi^*}(i) = \max_{\pi} \phi_{\pi}(i), \qquad \text{for all } i \geq 0,$$

and π^* is any policy that, for each i, prescribes an action that maximizes the right side of (2.1).

Proof: Let $H_t = (X_0, a_0, \ldots, X_t, a_t)$ denote the history of the process up to time t. Because $E(h(X_t)) = E\{E[h(X_t)|H_{t-1}]\}$, it follows that, for any policy π,

$$E_\pi \left\{ \sum_{t=1}^{n} \left[h(X_t) - E_\pi(h(X_t)|H_{t-1}) \right] \right\} = 0.$$

But,

$$E_\pi[h(X_t)|H_{t-1}] = \sum_{j=0}^{\infty} h(j) P_{X_{t-1} j}(a_{t-1})$$

$$= R(X_{t-1}, a_{t-1}) + \sum_{j=0}^{\infty} h(j) P_{X_{t-1} j}(a_{t-1})$$

$$- R(X_{t-1}, a_{t-1})$$

$$\leq \max_a \left[R(X_{t-1}, a) + \sum_{j=0}^{\infty} h(j) P_{X_{t-1} j}(a) \right]$$

$$- R(X_{t-1}, a_{t-1})$$

$$= g + h(X_{t-1}) - R(X_{t-1}, a_{t-1}),$$

with equality for π^* because π^* is defined to take the maximizing action. Hence,

$$0 \geq E_\pi \left\{ \sum_{t=1}^{n} \left[h(X_t) - g - h(X_{t-1}) + R(X_{t-1}, a_{t-1}) \right] \right\}$$

or

$$g \geq E_\pi \frac{[h(X_n)]}{n} - E_\pi \frac{[h(X_0)]}{n} + E_\pi \frac{\left[\sum_{t=1}^{n} R(X_{t-1}, a_{t-1}) \right]}{n}.$$

with equality for π^*. Letting $n \to \infty$ and using the fact that h is bounded, we have that

$$g \geq \phi_\pi(X_0),$$

with equality for π^* and for all possible values of X_0. Hence, the desired result is proven. $\square$

Therefore, if the conditions of Theorem 2.1 are satisfied, then a stationary optimal policy exists and may be characterized by the functional equation (2.1). This, however, immediately raises two questions. The first is why such a theorem should indeed be true. As it now stands, Theorem 2.1 seems in no way obvious or intuitive. Second, when are the conditions of Theorem 2.1 satisfied? In what follows, we shall attempt both to consider Theorem 2.1 intuitively and also to determine sufficient conditions for it to be of use.

Let us start by recalling the discounted-return criterion in which all future periods are discounted at rate α. In the average-return criterion all periods are given equal weight. Hence, it seems reasonable that under certain conditions the average return case should be obtainable as some type of limit of the discounted case as the discount factor α approaches unity.

Now, if we denote by V_α the optimal expected α-discounted return function, then V_α satisfies the optimality equation

$$V_\alpha(i) = \max_a \left[R(i, a) + \alpha \sum_j P_{ij}(a) V_\alpha(j) \right], \qquad i \geq 0, \qquad (2.2)$$

and the α-optimal policy selects the maximizing actions. One possible means of obtaining an optimal average-cost policy might be to choose the actions maximizing

$$\lim_{\alpha \to 1} \left[R(i, a) + \alpha \sum_{j=0}^{\infty} P_{ij}(a) V_\alpha(j) \right].$$

However, this limit need not exist and indeed would often be infinite for all actions. As a result, this direct approach is not fruitful.

Let us now consider the following indirect approach to the problem. Fix some state, say state 0, and let us define

$$h_\alpha(i) = V_\alpha(i) - V_\alpha(0)$$

to be the α-return of state i relative to state 0. Then, from (2.2), we obtain

$$(1 - \alpha)V_\alpha(0) + h_\alpha(i) = \max_a \left[R(i, a) + \alpha \sum_{j=0}^{\infty} P_{ij}(a)h_\alpha(j) \right]. \quad (2.3)$$

Also, it follows that the policy that chooses actions to maximize the right side of (2.3) is an α-optimal policy. Now, if for some sequence $\alpha_n \to 1$, $h_{\alpha_n}(j)$ converges to a function $h(j)$, and $(1 - \alpha_n)V_{\alpha_n}(0)$ converges to a constant g, then we obtain from (2.3)

$$g + h(i) = \max_a \left[R(i, a) + \sum_{j=0}^{\infty} P_{ij}(a)h(j) \right],$$

where we have assumed that the interchange of summation and limit is justified. Also, it seems reasonable that the policy that chooses the maximizing actions will be average-reward optimal, that is, that Theorem 2.1 will be true.

Theorem 2.2 If there exists an $N < \infty$ such that

$$|V_\alpha(i) - V_\alpha(0)| < N \qquad \text{for all } \alpha \text{ and all } i,$$

then

(i) there exists a bounded function $h(i)$ and a constant g satisfying (2.1);

(ii) for some sequence $\alpha_n \to 1$, $h(i) = \lim_{n \to \infty} [V_{\alpha_n}(i) - V_{\alpha_n}(0)]$;

(iii) $\lim_{\alpha \to 1}(1 - \alpha)V_\alpha(0) = g$.

Proof: With $h_\alpha(i) = V_\alpha(i) - V_\alpha(0)$, if follows by assumption that $h_\alpha(i)$ is uniformly bounded in i and α. Because every bounded sequence contains a convergent subsequence, it follows that there is a sequence $\alpha_{1,n} \to 1$ such that $\lim_{n \to \infty} h_{\alpha_{1,n}}(1) \equiv h(1)$ exists. Also, because $h_{\alpha_{1,n}}(2)$, $n \geq 1$, is bounded, it follows that there is a subsequence $\{\alpha_{2,n}\}$ of $\{\alpha_{1,n}\}$ such that $\lim_{n \to \infty} h_{\alpha_{2,n}}(2) \equiv h(2)$ exists. Similarly, there is a subsequence $\{\alpha_{3,n}\}$ of $\{\alpha_{2,n}\}$ for which $\lim_{n \to \infty} h_{\alpha_{3,n}}(3) \equiv h(3)$ exists, and so on. Now, if we let $\alpha_n = \alpha_{n,n}$, it is easy to see that $h_{\alpha_n}(i) \to h(i)$ as $n \to \infty$ for each i (why is this?). In addition, because rewards are bounded it follows that $(1 - \alpha_n)V_{\alpha_n}(0)$ is bounded, so there is a subsequence $\{\alpha_{\bar{n}}\}$ of $\{\alpha_n\}$ for which $\lim_{n \to \infty}(1 - \alpha_{\bar{n}})V_{\alpha_{\bar{n}}}(0) \equiv g$ exists.

Now, from (2.3) we have

$$(1 - \alpha_{\bar{n}})V_{\alpha_{\bar{n}}}(0) + h_{\alpha_{\bar{n}}}(i) = \max_a \left[R(i, a) + \alpha_{\bar{n}} \sum_{j=0}^{\infty} P_{ij}(a)h_{\alpha_{\bar{n}}}(j) \right].$$

Hence, part (i) follows by letting $n \to \infty$, noting that the boundedness of $h_{\alpha_{\bar{n}}}(j)$ implies, by the Lebesque-bounded convergence theorem, that

$$\sum_{j=0}^{\infty} P_{ij}(a)h_{\alpha_{\bar{n}}}(j) \to \sum_{j=0}^{\infty} P_{ij}(a)h(j) \qquad \text{as} \quad n \to \infty.$$

To prove (iii), we note that, because $(1 - \alpha)V_\alpha(0)$ is bounded, it follows that for any sequence $\alpha_n \to 1$ there is a subsequence $\{\alpha_{\bar{n}}, n \geq 1\}$ such that

$$\lim_{n \to \infty}(1 - \alpha_{\bar{n}})V_{\alpha_{\bar{n}}}(0)$$

exists. By the proof of (i) it follows that this limit must be g. Hence,

$$g = \lim_{\alpha \to 1}(1 - \alpha)V_\alpha(0),$$

and the proof is complete.   □

Remark:

(i)   From part (ii) of Theorem 2.2, it follows that $h(i)$ inherits the structural form of $V_\alpha(i)$. For instance, if $V_\alpha(i)$ is increasing (convex), then $h(i)$ will be increasing (convex).

(ii)   The technique used to show that if $f_\alpha(i)$ is uniformly bounded in $i = 0, 1, 2, \ldots$ and $\alpha \in (0, 1)$ then there exists a sequence $\alpha_n \to 1$ such that $\lim_{n \to \infty} f_{\alpha_n}(i)$ exists for each i is known as the *Cauchy diagonalization method*.

EXAMPLE 2.1a   Consider the machine-replacement example of Section II.3 (in which costs are being minimized rather than rewards maximized), and suppose (without loss of generality) that $C(i) \geq 0$ for all i. By the α-discounted optimality equation we have

$$V_\alpha(i) \leq C(i) + R + \alpha V_\alpha(0) \leq N + V_\alpha(0),$$

where N is such that $C(i) + R < N$. As a result, because $V_\alpha(i)$ was shown to be increasing in i, it follows that

$$|V_\alpha(i) - V_\alpha(0)| \leq N.$$

Hence, there exists a constant g and an increasing bounded function $h(i)$ such that

$$g + h(i) = C(i) + \min\left[R + h(0), \sum_j P_{ij}h(j)\right],$$

and the policy that chooses the minimizing actions is average-cost optimal. If we let

$$i^+ = \min\{i: \sum_j P_{ij}h(j) \geq R + h(0)\},$$

then it follows from the monotonicity in i of $\sum_j P_{ij}h(j)$ that the policy that replaces in all states greater than i^+ is average-cost optimal.

In determining a sufficient condition for $V_\alpha(i) - V_\alpha(0)$ to be uniformly bounded, we make use of the following result, known as Jensen's inequality, the proof of which will be left as an exercise.

Lemma 2.3 Jensen's Inequality For a convex function f and random variable X,

$$E[f(X)] \geq f(E[X]),$$

provided the expectations exist.

Theorem 2.4 If for some state, call it 0, there is a constant $N < \infty$ such that

$$m_{i0}(f_\alpha) < N \qquad \text{for all } i, \alpha,$$

then $V_\alpha(i) - V_\alpha(0)$ is uniformly bounded, where $m_{i0}(f_\alpha)$ is the expected time to go from state i to state 0 when using the α-discount optimal policy f_α.

Proof: Because the rewards are bounded and adding a constant to all rewards $R(i, a)$ will affect all policies identically, it follows that we may, without loss of generality, assume that all rewards are nonnegative. If we let

$$T = \min\{n: X_n = 0\},$$

then

$$V_\alpha(i) = E_{f_\alpha}\left[\sum_{n=0}^{T-1} \alpha^n R(X_n, a_n)\,\middle|\, X_0 = i\right] + E_{f_\alpha}\left[\sum_{n=T}^{\infty} \alpha^n R(X_n, a_n)\,\middle|\, X_0 = i\right]$$

$$\leq ME_{f_\alpha}(T) + V_\alpha(0)E_{f_\alpha}(\alpha^T)$$

$$\leq MN + V_\alpha(0), \tag{2.4}$$

where M is a bound on rewards. To obtain the inequality in the reverse direction, note by (2.4) that

$$V_\alpha(i) \geq V_\alpha(0)E_{f_\alpha}(\alpha^T)$$

or, equivalently,

$$V_\alpha(0) \leq V_\alpha(i) + \left[1 - E_{f_\alpha}(\alpha^T)\right]V_\alpha(0).$$

Now, $V_\alpha(0) \leq M/(1 - \alpha)$, and $E(\alpha^T) \geq \alpha^{E(T)} \geq \alpha^N$ by Jensen's inequality. Hence,

$$V_\alpha(0) \leq V_\alpha(i) + (1 - \alpha^N)\frac{M}{1 - \alpha}$$

$$\leq V_\alpha(i) + MN,$$

where the last inequality follows from

$$(1 - \alpha^N)/(1 - \alpha) = 1 + \alpha + \cdots + \alpha^{N-1} \leq N. \quad \square$$

As a corollary, we have the following.

Corollary 2.5   If the state space is finite and every stationary policy gives rise to an irreducible Markov chain, then $V_\alpha(i) - V_\alpha(0)$ is uniformly bounded, and hence the conditions of Theorem 2.2 are satisfied.

Proof: Because a finite-state Markov chain cannot be null recurrent, it follows that $m_{i0}(f) < \infty$ for all i and all stationary policies f. The result then follows from Theorem 2.4 because there are only a finite number of stationary policies. $\square$

Remark: Actually, it is not necessary that every stationary policy gives rise to an irreducible Markov chain. It is clearly sufficient if there is some state, say 0, that is accessible from every other state regardless of which α-optimal policy is being used.

3. Computational Approaches

3.1. Reduction to the Discounted Problem

In a special case, the average-return criterion may be reduced to a discounted-return criterion. We shall need the following assumption.

Assumption 3.1 There is a state, call it 0, and $\beta > 0$ such that

$$P_{i0}(a) \geq \beta \qquad \text{for all } i, \text{ all } a.$$

For any process satisfying this assumption, consider a new process with identical state and action spaces and identical rewards but with transition probabilities now given by

$$\bar{P}_{ij}(a) = \begin{cases} \dfrac{P_{ij}(a)}{1 - \beta}, & j \neq 0, \\[3mm] \dfrac{P_{i0}(a) - \beta}{1 - \beta}, & j = 0. \end{cases} \qquad (3.1)$$

Also, let $\bar{V}(i)$ be the $(1 - \beta)$-optimal value function for this new process. Now, letting

$$h(i) = \bar{V}(i) - \bar{V}(0),$$

we obtain, as in (2.3),

$$\beta \bar{V}(0) + h(i) = \max_a \Big[R(i, a) + (1 - \beta) \sum_j \bar{P}_{ij}(a) h(j) \Big]$$

$$= \max_a \Big[R(i, a) + \sum_j P_{ij}(a) h(j) \Big],$$

where the last equation follows from (3.1) because $h(0) = 0$. Hence, the conditions of Theorem 2.1 are satisfied. It follows that $g = \beta \bar{V}(0)$, and the average-return optimal policy is the one that selects the actions that maximize the right side of (3.2). But it is easily seen that this is precisely the $(1 - \beta)$-optimal policy for the new process. Hence, summing up, the average-return optimal policy is precisely the $(1 - \beta)$-optimal policy with respect to the new process; and the optimal average cost is $\beta \bar{V}(0)$.

Thus if our assumption holds, then we have reduced the average problem to a discounted-cost problem, and the methods of policy improvement or successive approximations may be employed.

3.2. A Linear Programming Solution

Let us suppose that our state space is finite, say $1, \ldots, M$, and that, under any stationary policy, the resultant Markov chain is ergodic. In this section we shall allow our stationary policies to be "randomized" in that their instructions may be to choose actions according to a probability distribution. In other words, a policy β is a set of numbers $\beta = \{\beta_i(a) : a \in A, \ i = 1, \ \ldots, \ M\}$ with the interpretation that, if the process is in state i, then action a is to be chosen with probability $\beta_i(a)$. Of course, we require that

$$0 \le \beta_i(a) \le 1, \qquad \text{for all } i, a,$$

$$\sum_a \beta_i(a) = 1, \qquad \text{for all } i.$$

Under any given policy β, the sequence of states $\{X_n : n = 0, 1, \ldots\}$ constitutes a Markov chain with transition probabilities $P_{ij}(\beta)$ given by†

$$P_{ij}(\beta) = P_\beta\{X_{n+1} = j \mid X_n = i\}$$

$$= \sum_a P_{ij}(a)\beta_i(a).$$

For any policy β, let π_{ia} denote the limiting (or steady-state) probability that the process will be in state i and action a will be chosen if policy β is employed. That is,

$$\pi_{ia} = \lim_{n \to \infty} P_\beta(X_n = i, a_n = a).$$

The vector $\pi = (\pi_{ia})$ must satisfy

(i) $\pi_{ia} \ge 0$ for all i, a.

(ii) $\sum_i \sum_a \pi_{ia} = 1.$

(iii) $\sum_a \pi_{ja} = \sum_i \sum_a \pi_{ia} P_{ij}(a)$ for all j.

(3.3)

Equations (3.3i) and (3.3ii) are obvious, and (3.3iii) follows because the left-hand side equals the steady-state probability of being in state j and the right-hand side is the same probability computed by conditioning on the state and action chosen one stage earlier.

Thus for any policy β, there is a vector $\pi = (\pi_{ia})$ that satisfies (3.3i)–(3.3iii), and with the interpretation that π_{ia} is equal to the steady-state probability of being in state i and choosing action a when policy β is employed. Moreover, it turns out that the reverse is also true. Namely, for any vector $\pi = (\pi_{ia})$ satisfying (3.3i)–(3.3iii), there exists a policy β such that if β is used, then the steady-state probability of being in i and

---

† We use the notation P_β to signify that the probability is conditional on the fact that policy β is used.

choosing action a equals π_{ia}. To verify this last statement, suppose that $\pi = (\pi_{ia})$ is a vector that satisfies (3.3i)–(3.3iii). Then, let the policy $\beta = \beta_i(a)$ be

$$\beta_i(a) = P(\beta \text{ chooses } a \,|\, \text{state is } i)$$

$$= \frac{\pi_{ia}}{\sum_a \pi_{ia}}.$$

Now, let P_{ia} denote the limiting probability of being in i and choosing a when policy β is employed. We must show that $P_{ia} = \pi_{ia}$. To do so, first note that $\{P_{ia}, i = 1, \ldots, M, a \in A\}$ are the limiting probabilities of the two-dimensional Markov chain $\{(X_n, a_n), n \geq 0\}$. Hence, they are the unique solution of

$$\text{(i')} \quad P_{ia} \geq 0,$$

$$\text{(ii')} \quad \sum_i \sum_a P_{ia} = 1, \tag{3.3'}$$

$$\text{(iii')} \quad P_{ja} = \sum_i \sum_{a'} P_{ia'} P_{ij}(a')\beta_j(a),$$

where (3.3iii') follows because

$$P(X_{n+1} = j, a_{n+1} = a \,|\, X_n = i, a_n = a') = P_{ij}(a')\beta_j(a).$$

Because

$$\beta_j(a) = \frac{\pi_{ja}}{\sum_a \pi_{ja}},$$

we see that (P_{ia}) is the unique solution of

$$P_{ia} \geq 0$$

$$\sum_i \sum_a P_{ia} = 1$$

$$P_{ja} = \sum_i \sum_{a'} P_{ia'} P_{ij}(a') \frac{\pi_{ja}}{\sum_a \pi_{ja}}.$$

Hence, to show that $P_{ia} = \pi_{ia}$, we must show that

$$\pi_{ia} \geq 0$$

$$\sum_i \sum_a \pi_{ia} = 1$$

$$\pi_{ja} = \sum_i \sum_{a'} \pi_{ia'} P_{ij}(a') \frac{\pi_{ja}}{\sum_a \pi_{ja}}.$$

The first two equations follow from (3.3i) and (3.3ii) and the third, which is equivalent to

$$\sum_a \pi_{ja} = \sum_i \sum_{a'} \pi_{ia'} P_{ij}(a'),$$

follows from condition (3.3iii).

Thus we have shown that a vector $\pi = (\pi_{ia})$ will satisfy (3.3i), (3.3ii), and (3.3iii) if and only if there exists a policy β such that π_{ia} is equal to the steady-state probability of being in state i and choosing action a when β is used. In fact, the policy β is defined by

$$\beta_i(a) = \frac{\pi_{ia}}{\sum_a \pi_{ia}}.$$

Now,

$$\text{expected average reward under } \beta = \lim_{n \to \infty} E_\beta \left[\frac{\sum_{i=1}^n R(X_i, a_i)}{n} \right].$$

If π_{ia} denotes the steady-state probability of being in state i and choosing action a, it follows that the limiting expected reward at time n equals

$$\lim_{n \to \infty} E(R(X_n, a_n)) = \sum_i \sum_a \pi_{ia} R(i, a),$$

which implies that

$$\text{expected average reward under } \beta = \sum_i \sum_a \pi_{ia} R(i, a).$$

Hence, the problem of determining the policy that maximizes the expected average reward is

$$\text{maximize} \quad \sum_i \sum_a \pi_{ia} R(i, a), \tag{3.4}$$

$$\text{subject to} \quad \pi_{ia} \geq 0, \quad \text{for all } i, a,$$

$$\sum_i \sum_a \pi_{ia} = 1$$

$$\sum_a \pi_{ja} = \sum_i \sum_a \pi_{ia} P_{ij}(a) \quad \text{for all } j.$$

However, this maximization problem is a special case of what is known as a *linear program*† and can thus be solved by a standard linear

---

† It is called a linear program because the objective function $\sum_i \sum_a R(i, a)\pi_{ia}$ and the constraints are all linear functions of the π_{ia}.

programming algorithm known as the *simplex algorithm.* If $\pi^* = (\pi_{ia}^*)$ maximizes (3.4), then the optimal policy will be given by β^*, where

$$\beta_i^*(a) = \frac{\pi_{ia}^*}{\sum_a \pi_{ia}^*}.$$

Remarks:

(i) The π^* maximizing Eq. (3.4) has, of course, the property that, for each i, π_{ia}^* is zero for all but one value of a, which implies that the optimal policy is nonrandomized, that is, the action it prescribes when in state i is a deterministic function of i.

(ii) The linear programming formulation often works when there are restrictions placed on the class of allowable policies. For instance, suppose there is a restriction on the fraction of time the process spends in some state, say, state 1. Specifically, suppose that we are allowed to consider only policies having the property that their use results in the process being in state 1 less than 100α percent of the time. To determine the optimal policy subject to this requirement, we add to the linear programming problem the additional constraint

$$\sum_a \pi_{1a} \leq \alpha,$$

because $\sum_a \pi_{1a}$ represents the proportion of time that the process is in state 1.

(iii) Whereas we have assumed that each stationary policy gives rise to an ergodic Markov chain, the results go through in identical fashion as long as the resulting chain is irreducible (even though it may be periodic). This is seen by replacing "limiting" probabilities by "stationary" probabilities throughout.

Problems

A randomized stationary policy is one whose decision depends only on the present state but is allowed to be randomized in that it instructs the decision maker to select actions according to a probability distribution. That is, a randomized stationary policy is $\pi = \{\pi(i, a)\}$, with the instruction that when in state i action a is chosen with probability $\pi(i, a)$ where

$$\pi(i, a) \geq 0, \qquad \sum_a \pi(i, a) = 1.$$

1.   For Counterexample 1.2, consider the randomized stationary policy for which

$$\pi(i, a_2) = \frac{i}{i + 1}, \qquad \pi(i, a_1) = \frac{1}{i + 1}.$$

Show that the average return under π is 1.

2.   For Counterexample 1.2, show that the average return for any randomized stationary policy is 0.

3.   A quality control model: Consider a machine that can be in one of two states, good or bad. Suppose that the machine produces an item at the beginning of each day. The item produced is either good (if the machine is good) or bad (if the machine is bad). Suppose that once the machine is in the bad state, it remains in that state until it is replaced. However, if it is in the good state at the beginning of a day, then with probability γ it will be in the bad state at the beginning of the next day.

    We further suppose that after the item is produced we have the option of inspecting the item or not. If the item is inspected and found to be in the bad state, then the machine is immediately replaced with a good machine at an additional cost R. Also, the cost of inspecting an item will be denoted by I, and the cost of producing a bad item by C.

    Say that the process is in state p at time t if p is the posterior probability at t that the machine in use is in the bad state. If the objective is to minimize the average expected cost per unit time, determine the structure of the optimal policy.

4.   Consider a machine that can be in any one of two states, good or bad. At the beginning of each day the machine produces items that are either defective or nondefective. The probability of a defective item is P_1, when in the good state and P_2 when in the bad state. Once in the bad state, the machine remains in this state until it is replaced. However, if the machine is in the good state at the beginning of one day, then with probability γ it will be in the bad state at the beginning of the next day. A decision as to whether or not to replace the machine must be made each day after observing the item produced. Let R be the cost of replacing the machine and let C be the cost incurred whenever a defective item is produced. Prove that, for the average-cost optimal policy, there is a p^* such that the optimal policy replaces the machine whenever the present

probability that the process is in the bad state is greater than or equal to p^*.

5. We say that the stationary policies f_n converge to the stationary policy f, and write $f_n \rightarrow f$, if for each i there exists an N_i such that $f_n(i) = f(i)$ for all $n \geq N_i$. Show that if $V_\alpha(i) - V_\alpha(0)$ is uniformly bounded in α and i, then for some sequence $\alpha_n \rightarrow 1$ and some average cost optimal policy f^*, $f_{\alpha_n} \rightarrow f^*$.

6. Suppose that $V_\alpha(i) - V_\alpha(0)$ is uniformly bounded. Show that, if $f_{\alpha_n} \rightarrow f$ for some sequence $\alpha_n \rightarrow 1$, then f is average-cost optimal.

7. Suppose that $P_{i0}(a) = \beta$ for all i and a. Show that for any stationary policy f, $\phi_f(0) = \beta \bar{V}_f(0)$, where $\bar{V}_f(0)$ is the expected $(1 - \beta)$-discount cost when the transition probabilities are given by (3.1).

8. Argue that the result of Problem 7 is true whenever $P_{i0}(a) \geq \beta$ for all i and a.

9. If $V_\alpha(i) - V_\alpha(0)$ is uniformly bounded, show that

$$(1 - \alpha)V_\alpha(j) \rightarrow g \qquad \text{for all } j.$$

10. Prove Jensen's inequality.

11. Use the duality theory of linear programming to prove that the optimal solution π^* of (3.4) can be chosen so that π_{ia}^* is, for each i, zero for all but one value of a.

Notes and References

Counterexamples 2 and 3 come from Ross [4] and [5]. The approach of this chapter is taken from Ross [3]. For a different approach the reader should see Derman [1]. The first person to have noticed that linear programming may be used to solve the average-cost model seems to have been Manne [2]. For an elementary introduction to Markov chains, the interested reader is referred to Ross [6].

1. Derman, C., "Denumberable state Markovian decision processes—average cost criterion," *Ann. Math. Statist.* **37**, 1545–1554, 1966.

2. Manne, A., "Linear programming and sequential decisions," *Management Sci.* **6** (3), 259–267, 1960.

3. Ross, S. M., "Non-discounted denumerable Markovian decision models," *Ann. Math. Statist.* **39** (2), 412–423, 1968.

4. Ross, S. M., *Applied Probability Models with Optimization Applications*. Holden–Day, San Francisco, 1970.
5. Ross, S. M., "On the nonexistence of ϵ-optimal randomized stationary policies in average cost Markovian decision models," *Ann. Math. Statist.* **42** (5), 1767–1768, 1971.
6. Ross, S. M., *Introduction to Probability Models*, 2nd ed. Academic Press, New York, 1980.

VI

Stochastic Scheduling

1. Introduction

In this chapter we deal with some problems of scheduling, assigning, and ordering. We start in Section 2 with a model for maximizing the expected reward earned in a fixed time. Specifically, we suppose that there are n jobs to be processed in sequential order, with the ith job taking an exponentially distributed length of time with rate λ_i to complete and earning the decision maker R_i upon completion. We show that if the jobs are processed in decreasing order of $\lambda_i R_i$, $i = 1, \ldots, n$, then the expected reward earned by t is maximized for all t.

In Sections 3 and 4 we consider problems consisting of 2 processors and n jobs needing to be processed. In both models we take as our objective the minimization of the expected time until all work is completed. In Section 3 we suppose that the processing time of job i, on either machine, is exponentially distributed with rate u_i and that each job must be processed on either machine. In Section 4 we suppose that each job must be processed first on processor I and then on processor II, and assuming that the processing time of a given job on a given processor is exponential, we determine the optimal policy.

In Section 5 we consider a set of n items, where the ith item has a rating X_i. These items are to be released to the field in a sequential fashion. That is, the ith-issued item is to be placed in field immediately upon the death of the $(i - 1)$st-issued item. It is supposed that an item with rating X will have a field life of $X d(t)$ if released at time t, where $d(t)$ is an arbitrary deterministic function. The problem of interest is to determine the optimal order to issue the items so as to maximize the total expected field life.

In Sections 6 and 7 we consider stochastic versions of two famous deterministic optimization problems; the knapsack problem and the assignment problem.

2. Maximizing Finite-Time Returns— Single Processor

Suppose there are n jobs to be performed sequentially within a fixed time t. The ith job takes an exponential amount of time with mean $1/\lambda_i$ and, if completed within the time of the problem, earns the decision maker an amount R_i. Whenever a job is completed, the decision maker must decide which of the remaining jobs to attempt, with his objective being to maximize the total expected earnings.

It follows from the lack-of-memory property of the exponential that, if job i is attempted for a time dt, then it will be completed with probability $\lambda_i \, dt + o(dt)$, thus the expected gain will be $\lambda_i R_i dt + o(dt)$. Hence, it seems as if the expected return is the same as if we earned $\lambda_i R_i$ per unit time that job i is being performed. To show this formally, suppose that t units of time remain when job i is initiated. If X_i is the time needed to perform this job, then the expected return from job i is

$$E(\text{return from } i) = R_i P(X_i < t)$$

$$= R_i(1 - e^{-\lambda_i t})$$

$$= \lambda_i R_i \frac{1 - e^{-\lambda_i t}}{\lambda_i}$$

$$= \lambda_i R_i E[\min(X_i, t)]$$

$$= \lambda_i R_i E(\text{length of time } i \text{ is worked on}),$$

where the identity

$$E[\min(X_i, t)] = \int_0^\infty P[\min(X_i, t) > x] \, dx$$

$$= \int_0^t e^{-\lambda_i x} \, dx = (1 - e^{-\lambda_i t})/\lambda_i$$

has been used. Hence, it follows that, for any policy π,

$$E_\pi(\text{total return}) = \sum_{i=1}^{n} \lambda_i R_i E_\pi(\text{time } i \text{ is performed}). \qquad (2.1)$$

That is, the total expected return is the same as it would be if we earned money at a rate $\lambda_i R_i$ whenever job i was being worked on (whether or not it is ever completed). From this we see that the expected amount earned by time t is maximized by working on jobs in decreasing order of $\lambda_i R_i$. That is, we have shown the following proposition.

Proposition 2.1 The expected return by time t is maximized by attempting jobs in decreasing order of $\lambda_i R_i$, $i = 1, \ldots, n$.

3. Minimizing Expected Makespan— Processors in Parallel

Suppose that two identical parallel processors are available to perform n tasks. To perform task j, that task has to be put on one of the processors (either of the two can do) for a duration X_j, after which it is complete. Tasks are performed consecutively starting at time $t = 0$, so that as soon as a task is complete another task is put on the processor that is freed. Suppose that X_j, the time to process task j, is exponentially distributed with rate u_j, $j = 1, \ldots, n$, independently of all that has occurred prior to it being put on one of the processors. For any permutation $i_1, \ldots, i_n$ of $1, 2, \ldots, n$, putting tasks on the processors in that order defines a schedule. The time until all jobs have been processed is called the *makespan*, and our objective is to determine the schedule that minimizes the expected makespan.

It will help our analysis to assume that at time 0 one of the processors is occupied on job 0 and will remain occupied for a time X_0, where X_0 is assumed to have an arbitrary distribution and is independent of $X_1, \ldots, X_n$.

For any schedule $\pi = (0, i_1, \ldots, i_n)$, let M denote the makespan and D the amount of time that only one of the processors is busy. That is, at time $M - D$ one of the processors completes work on a job and finds no other jobs available. As the total amount of work processed is $M + M - D$, which must equal $\sum_0^n X_j$, we see that

$$2M = D + \sum_0^n X_j,$$

so, for any strategy π,

$$2E_\pi(M) = E_\pi(D) + \sum_0^n E(X_j). \qquad (3.1)$$

Thus minimizing the expected difference of the times at which the processors become idle also leads to minimizing the expected makespan.

The following lemma will be used to show that scheduling jobs in increasing order of their service rates is optimal.

Lemma 3.1   Consider the policies $\pi = (0, 2, 1, 3, \ldots, n)$ and $\bar\pi = (0, 1, 2, 3, \ldots, n)$. If $u_1 = \min_{k \geq 1} u_k$, then $E_{\bar\pi}(D) \leq E_\pi(D)$.

Proof: Let P_j and $\bar P_j$, $j \geq 0$, be the probabilities that the last job to be completed is job j, under policies π and $\bar\pi$, respectively. Clearly $P_0 = \bar P_0 = P(X_0 > \sum_1^n X_j)$, and we shall prove by induction that

$$\bar P_1 \leq P_1, \qquad \bar P_j \geq P_j, \qquad j = 2, \ldots, n. \qquad (3.2)$$

This is obvious when $n = 1$, so assume it to be true whenever there are only $n - 1$ jobs (in addition to job 0) to be scheduled. Now consider the n-job case, and let P_j^* and $\bar P_j^*$, $j = 1, \ldots, n - 1$, be the probabilities that job j is the last of jobs $0, 1, \ldots, n - 1$ to be completed under policies π and $\bar\pi$, respectively. Then, by the induction hypothesis,

$$\bar P_1^* \leq P_1^*, \qquad \bar P_j^* \geq P_j^*, \qquad j = 2, \ldots, n - 1. \qquad (3.3)$$

However, using the lack of memory of the exponential and the fact that job n is the last to begin processing under both π and $\bar\pi$, we have

$$P_j = P_j^* \frac{u_n}{u_n + u_j}, \qquad \bar P_j = \bar P_j^* \frac{u_n}{u_n + u_j}, \qquad j = 1, \ldots, n - 1.$$

Hence, from the induction hypothesis (3.3),

$$\bar P_1 \leq P_1, \qquad \bar P_j \geq P_j, \qquad j = 2, \ldots, n - 1.$$

Also,

$$\bar{P}_n - P_n = \sum_{j=0}^{n-1} (\bar{P}_j^* - P_j^*) \frac{u_j}{u_j + u_n}$$

$$= (\bar{P}_1^* - P_1^*) \frac{u_1}{u_1 + u_n} + \sum_{j=2}^{n-1} (\bar{P}_j^* - P_j^*) \frac{u_j}{u_j + u_n}$$

$$\geq \frac{u_1}{u_1 + u_n} \sum_{j=1}^{n-1} (\bar{P}_j^* - P_j^*)$$

$$= 0,$$

where the last inequality follows because $u_j \geq u_1$ implies that

$$\frac{u_j}{u_j + u_n} \geq \frac{u_1}{u_1 + u_n}.$$

Hence, the induction proof is now complete. □

Because

$$E_\pi(D) = \sum_{j=1}^{n} \frac{1}{u_j} P_j + P_0 E\left(X_0 - \sum_{1}^{n} X_i \mid X_0 > \sum_{1}^{n} X_i\right),$$

we see that

$$E_\pi(D) - E_{\bar{\pi}}(D) = \sum_{j=1}^{n} \frac{1}{u_j} (P_j - \bar{P}_j)$$

$$\geq \frac{1}{u_1} (P_1 - \bar{P}_1) + \frac{1}{u_1} \sum_{j=2}^{n} (P_j - \bar{P}_j)$$

$$= 0,$$

where the inequality follows from (3.2) because $u_1 = \min_k u_k$, and the final equality follows from $\sum_1^n P_j = \sum_1^n \bar{P}_j$.

Theorem 3.2 If $u_1 \leq u_2 \leq \cdots \leq u_n$, then the expected makespan is minimized under strategy $(0, 1, 2, \ldots, n)$.

Proof: Consider any arbitrary strategy that does not initially process job 1, say, $(0, i_1, \ldots, i_k, i_{k+1}, 1, \ldots)$. By considering this at the time at which only one of the jobs, $0, i_1, \ldots, i_k$, have not yet finished its processing, we see, using Lemma 3.1, that the schedule $(0, i_1, \ldots, i_k, 1, i_{k+1}, \ldots)$ has a smaller expected makespan. Continuing with this we see that $(0, 1, i_1, \ldots, i_{k+1}, \ldots)$ is better. If $i_1 \neq 2$, then, repeating the preceding argument, we show that $(0, 1, 2, i_1, \ldots)$ is better. Continuing in this manner shows that the strategy $(0, 1, 2, \ldots, n)$ minimizes $E(D)$, which, using (3.1), proves the result. $\square$

Remarks:

(i)   Whereas Theorem 3.2 only proved that scheduling tasks in increasing order of their exponential service rates is optimal among the $n!$ strategies that determine their ordering in advance, it is also optimal among all possible policies. That is, it remains optimal even when the choice of task to begin processing is allowed to depend on what has occurred up to that time. This is easily shown by induction as follows: It is immediate when $n = 1$, so assume it to be true whenever there are $n - 1$ tasks to be processed. Now, whichever of the n tasks is initially processed (alongside task 0), at the moment one of the two processors become free, it follows by the induction hypothesis that the remaining tasks should be scheduled in increasing order of their rates. Hence, the only policies we need consider are those n for which task $i, i = 1, \ldots, n$, is scheduled first, and the remaining tasks are scheduled in increasing order of their rates. But Theorem 3.2 shows that the optimal policy of this type is the one that schedules the n tasks in increasing order of their rates. This completes the induction.

(ii)   Theorem 3.2 can be shown to remain valid when there are more than two processors available. The proof, however, is more involved.

(iii)   Not all pairwise switches from $(\ldots, 2, 1, i_k, \ldots, i_n)$ to $(\ldots, 1, 2, i_k, \ldots, i_n)$ are beneficial when $u_1 \leq u_2$. We also require that u_1 be smaller than the service rate of all jobs following it, that is, $u_1 \leq \min(u_{i_k}, \ldots, u_{i_n})$.

(iv)   It can be shown that the policy given in Theorem 3.2 has the property that it stochastically minimizes the makespan. That is, for any $a \geq 0$, the probability that the makespan exceeds a is minimized by this policy. This is a stronger result than that given by Theorem 3.2, which states only that the policy minimizes the expected makespan. In addition, it can also be shown that the stated policy stochastically maximizes the time until one of the processors becomes idle. That is, in the notation of this section, it maximizes the probability that $M - D$ exceeds a for each $a \geq 0$.

Let us now suppose that a return R_j is earned when task j is performed and our objective is to maximize the expected total return earned by some fixed time t. It follows, as in (2.1), that the total expected return under any policy π can be expressed as

$$E_\pi(\text{total return by } t) = \sum_j u_j R_j E_\pi(\text{processing time of task } j \text{ by } t).$$

Thus, at first glance, it might seem that the optimal policy would be to sequence the tasks in decreasing order of $u_j R_j$ (as in the case when there is only a single processor). To see that this need not be the case, suppose that $u_j R_j \equiv 1$ and $u_1 < u_2 < \cdots < u_n$. Hence, the expected return by time t for any policy is equal to the expected total processing time on all tasks by t. Because the policy given in Theorem 3.2 uniquely stochastically maximizes the time until one of the processors becomes idle, it also uniquely stochastically maximizes the total processing time by t and is thus uniquely optimal under our new objective function. However, this contradicts the conjecture that it is optimal to process tasks in decreasing order of $u_j R_j$, for, in the case $u_j R_j \equiv 1$, the conjecture would imply that all orderings are optimal. We can, however, prove that this policy is optimal in a special case.

Theorem 3.3   If

$$u_1 \leq u_2 \leq \cdots \leq u_n,$$

and

$$u_1 R_1 \geq u_2 R_2 \geq \cdots \geq u_n R_n \geq 0,$$

then sequencing the tasks in the order $1, 2, \ldots, n$ maximizes the expected return earned by t for each $t > 0$.

Proof: Fix t and let T_j denote the expected total processing time of task j by time t. Now, because the policy that sequences according to $1, 2, \ldots, n$ stochastically maximizes the time until one of the processors becomes idle, it easily follows that it also stochastically maximizes the total processing time by t. (Prove this.) Because this remains true even when the set of tasks is $1, 2, \ldots, j$, it follows that $\sum_{i=1}^j T_i$ is, for each $j = 1, \ldots, n$, maximized by the policy under consideration. The result now follows from the following lemma, with $c(i) = u_i R_i$.

Lemma 3.4   If $T_1, \ldots, T_n$ and $S_1, \ldots, S_n$ are nonnegative numbers such that

$$\sum_{i=1}^j T_i \geq \sum_{i=1}^j S_i, \qquad j = 1, \ldots, n,$$

and if $c(1) \geq c(2) \geq \cdots \geq c(n) \geq 0$, then

$$\sum_{i=1}^{n} c(i)T_i \geq \sum_{i=1}^{n} c(i)S_i.$$

Proof: Let

$$T_{n+1} = 0,$$

$$T = \sum_{i=1}^{n+1} T_i,$$

$$S_{n+1} = T - \sum_{i=1}^{n} S_i \geq 0.$$

Also, let X and Y be random variables such that

$$P(X = i) = T_i/T, \qquad i = 1, \ldots, n+1,$$

$$P(Y = i) = S_i/T, \qquad i = 1, \ldots, n+1.$$

Now the hypothesis of the lemma states that

$$X \underset{\mathrm{st}}{\leq} Y.$$

Because $c(j)$ is a decreasing function, $j = 1, \ldots, n+1$, where $c(n+1) \equiv 0$, it thus follows that

$$E(c(X)) \geq E(c(Y)),$$

which proves the result. $\square$

4. Minimizing Expected Makespan— Processors in Series

Suppose that there are n jobs and two machines. Each of the jobs must be processed on both machines, first on machine A and then on machine B. The processing times for job j on machines A and B are exponential random variables with rates λ_j and μ_j, respectively. Once a job has finished being processed at machine A, it then joins the queue at machine B. The objective is to determine the order in which to process jobs at machine A to minimize the expected time until all jobs have been processed on both machines. (Clearly the order in which jobs are processed at machine B is irrelevant as long as that machine is never kept idle while jobs are waiting in its queue.)

To gain some insight, let us consider the case in which $n = 2$. If job 1 is processed first on machine A, then the expected completion time, call it $E(C_{1,2})$, is given by

$$E(C_{1,2}) = \frac{1}{\lambda_1} + \frac{1}{\mu_1 + \lambda_2} + \frac{\mu_1}{\mu_1 + \lambda_2}\left(\frac{1}{\lambda_2} + \frac{1}{\mu_2}\right) + \frac{\lambda_2}{\mu_1 + \lambda_2}\left(\frac{1}{\mu_1} + \frac{1}{\mu_2}\right).$$

This follows because $1/\lambda_1$ is the mean time until job 1 is completed on machine A, at which time job 1 goes to machine B and job 2 goes to A. Then $1/(\mu_1 + \lambda_2)$ is the mean time until either job 1 is completed at B or job 2 at A. The other terms are then obtained by conditioning on whichever occurs first (either 1 finishing at B or 2 at A).

Similarly, by reversing the order we have that $E(C_{2,1})$ is given by

$$E(C_{2,1}) = \frac{1}{\lambda_2} + \frac{1}{\mu_2 + \lambda_1} + \frac{\mu_2}{\mu_2 + \lambda_1}\left(\frac{1}{\lambda_1} + \frac{1}{\mu_1}\right) + \frac{\lambda_1}{\mu_2 + \lambda_1}\left(\frac{1}{\mu_1} + \frac{1}{\mu_2}\right).$$

Simple algebra now shows that

$$E(C_{1,2}) \leq E(C_{2,1}) \Leftrightarrow \lambda_1 - \mu_1 \geq \lambda_2 - \mu_2.$$

That is, for each job, we consider the difference between that job's service rates at A and B and schedule first the one whose difference is largest. We now show that this remains true even when there are more than two jobs.

Let X_j denote the processing time for job j at machine A (so X_j is exponential with rate λ_j). If C is the completion time under any schedule, let

$$R = C - \sum_{j=1}^{n} X_j.$$

Thus R is the "remainder" time, that is, it represents the amount of work that remains at machine B when machine A has completed its processing. Hence,

$$E(R) = E(C) - \sum_{j=1}^{n} 1/\lambda_j,$$

so minimizing $E(C)$ is equivalent to minimizing $E(R)$. We shall prove that the policy that schedules jobs at machine A in decreasing order of

$\lambda_j - \mu_j$ minimizes $E(R)$. In fact, we shall use a "pairwise interchange" argument to show that this ordering stochastically minimizes R [and thus minimizes $E(R)$].

Consider first the case of $n = 2$ jobs, and suppose that, initially at time $t = 0$, machine B is occupied with the amount of work z. That is, B must spend z units working on prior work before it can start processing either job 1 or 2. Let $R_{1,2}(z)$ denote the remainder, that is, $R_{1,2}(z) = C - X_1 - X_2$, when job 1 is scheduled first, and similarly for $R_{2,1}(z)$.

The following lemma shows that the suggested ordering stochastically minimizes $R(z)$ for any z.

Lemma 4.1   If $\lambda_1 - \mu_1 \geq \lambda_2 - \mu_2$, then for any z

$$R_{1,2}(z) \underset{st}{\leq} R_{2,1}(z).$$

Proof: We have to compare $P[R_{1,2}(z) > a]$ with $P[R_{2,1}(z) > a]$. When $z \geq X_1 + X_2$, these terms are equal (because in either case $R = z - X_1 - X_2 + Y_1 + Y_2$, where Y_i is the processing time of job i at B). Hence, we need only look at $P[R_{1,2}(z) > a \mid X_1 + X_2 > z]$. Now, stating that $X_1 + X_2 > z$ is equivalent to stating that at some time job 1 will be in machine B and job 2 in machine A. Hence, using the lack of memory of the exponential, and conditioning on which machine finishes first, we see that

$$P[R_{1,2}(z) > a \mid X_1 + X_2 > z]$$

$$= \frac{\mu_1}{\mu_1 + \lambda_2} e^{-\mu_2 a} + \frac{\lambda_2}{\mu_1 + \lambda_2} P[e^{\mu_1} + e^{\mu_2} > a]$$

$$= \frac{\mu_1}{\mu_1 + \lambda_2} e^{-\mu_2 a} + \frac{\lambda_2}{\mu_1 + \lambda_2} \left[e^{-\mu_1 a} + \int_0^a \mu_1 e^{-\mu_1 x} e^{-\mu_2(a-x)} \, dx \right]$$

$$= \frac{\mu_1}{\mu_1 + \lambda_2} e^{-\mu_2 a} + \frac{\lambda_2(\mu_1 e^{-\mu_2 a} - \mu_2 e^{-\mu_1 a})}{(\mu_1 + \lambda_2)(\mu_1 - \mu_2)}$$

$$= \frac{\mu_1(\mu_1 - \mu_2 + \lambda_2)e^{-\mu_2 a} - \mu_2 \lambda_2 e^{-\mu_1 a}}{(\mu_1 + \lambda_2)(\mu_1 - \mu_2)}.$$

Because the expression for $P[R_{2,1}(z) > a \mid X_1 + X_2 > z]$ is similar, we see that

$$P[R_{2,1}(z) > a | X_1 + X_2 > z] - P[R_{1,2}(z) > a | X_1 + X_2 > z]$$

$$= \frac{\mu_1 \mu_2}{(\mu_1 + \lambda_2)(\mu_2 + \lambda_1)} \frac{e^{-\mu_1 a} - e^{-\mu_2 a}}{\mu_2 - \mu_1} [\lambda_1 - \mu_1 - (\lambda_2 - \mu_2)],$$

$$\geq 0$$

which completes the result. $\square$

Proposition 4.2   For any initial workload of machine B, R is stochastically minimized, and thus $E(C)$ is minimized, by scheduling jobs to be processed on A in decreasing order of $\lambda_j - \mu_j$.

Proof: Consider first any of the $n!$ policies in which the ordering is fixed at time 0. Suppose $\lambda_1 - \mu_1 = \max_j (\lambda_j - \mu_j)$ and that the ordering calls for job j to be processed on A immediately before job 1. Then at the moment at which machine A is to begin on job j, no matter what the remaining work is at machine B at that moment, it follows from Lemma 4.1 that, if we interchange jobs 1 and j, then the remaining work at machine B when both 1 and j have been processed at A will be stochastically reduced. But it is obvious that, for a given set of jobs to be processed in both machines, the remainder time is a stochastically increasing function of the initial workload of machine B. Hence, the remainder time is stochastically reduced by the interchange. Repeated use of this interchange argument shows that the suggested policy stochastically minimizes the remainder time among all the $n!$ policies whose ordering is fixed at time 0. Hence, it minimizes the expected remainder time, and thus the expected completion time, among all such policies.

To show that it is optimal among all policies follows by induction. It is immediate for $n = 1$, so assume it whenever there are $n - 1$ jobs to be processed on the two machines no matter what the initial workload of machine B. Now no matter which job is initially processed at machine A, at the moment its processing at A is completed, it follows by the induction hypothesis that the remaining jobs are processed in decreasing order of the difference of their rates at machines A and B. Hence, we need only consider fixed-order policies, and thus this policy is optimal. $\square$

5. Maximizing Total Field Life

Consider a stockpile consisting of n items, where the ith item has a rating X_i, $i = 1, \ldots n$. An item with rating X, if kept in stockpile until time t and then released to the field, will have a field life of $Xd(t)$. Thus it is assumed that, for any t, the field life for issuance at time t is proportional to the time-0 field life.

Items are to be issued one by one from the stockpile to the field until the stockpile is depleted. The ith issued item is placed in the field immediately upon the death in the field of the $(i - 1)$st issued item. The problem of interest is to find the order-of-item issue that maximizes the total field life obtained from the stockpile.

Suppose that the ratings X_i, $i = 1, \ldots, n$, are random variables that are likelihood-ratio ordered. That is, suppose that

$$\frac{f_i(x)}{f_{i+1}(x)} \quad \text{increases in } x, \qquad i = 1, \ldots, n - 1,$$

where f_i is the density function of X_i. We shall make use of the following useful result concerning such ordered random variables.

Recall that $X \geq_{st} Y$ means that

$$P[X > a] \geq P[Y > a] \qquad \text{for all } a,$$

whereas $X \geq_{LR} Y$ means that

$$\frac{f(x)}{g(x)} \quad \text{increases in } x,$$

where f and g are the densities of X and Y.

Proposition 5.1 Suppose that X and Y are independent, with densities f and g, respectively, and suppose that

$$Y \underset{LR}{\leq} X.$$

If $h(y, x)$ is a real-valued function satisfying

$$h(y, x) \geq h(x, y) \qquad \text{whenever } y \leq x,$$

then

$$h(Y, X) \underset{st}{\geq} h(X, Y).$$

Proof: Let $U = \max(X, Y)$, $V = \min(X, Y)$. Then, conditional on $U = u$, $V = v$, $u \geq v$, the conditional distribution of $h(X, Y)$ is concentrated on the two points $h(u, v)$ and $h(v, u)$, assigning probability

$$\lambda_1 \equiv P[X = \max(X, Y), Y = \min(X, Y) | U = u, V = v]$$

$$= \frac{f(u)g(v)}{f(u)g(v) + f(v)g(u)}$$

to the smaller value $h(u, v)$. Similarly, conditional on $U = u$ and $V = v$, $h(Y, X)$ is also concentrated on the two points $h(u, v)$ and $h(v, u)$, assigning probability

$$\lambda_2 \equiv P[Y = \max(X, Y), X = \min(X, Y) | U = u, V = v]$$

$$= \frac{g(u)f(v)}{g(u)f(v) + f(u)g(v)}$$

to the smaller value. As $u \geq v$, because $X \geq_{\text{LR}} Y$,

$$f(u)g(v) \geq g(u)f(v);$$

so, conditional on $U = u$ and $V = v$, $h(Y, X)$ is stochastically larger than $h(X, Y)$. That is,

$$P[h(Y, X) \geq a | U, V] \geq P[h(X, Y) \geq a | U, V],$$

and the result now follows by taking expectations of both sides. $\square$

We shall suppose for the remainder of this section that $d(t)$ is a nonnegative concave function. (It should be noted that this condition implies that d is an increasing function, which means that items are improving while in the stockpile.) The corresponding results for convex d will be left as exercises.

We start by showing that, if at any time there are only two more items to be issued, then the remaining field life is stochastically maximized by first issuing the item that is likelihood-ratio ordered smaller.

Lemma 5.2 If $0 < r_1 < r_2$ and $d(t)$ is a nonnegative concave function, then for all $y \geq 0$

$$r_1 d(y) + r_2 d(y + r_1 d(y)) \geq r_2 d(y) + r_1 d(y + r_2 d(y)).$$

Proof: Fix y. If $d(y) = 0$, then the preceding inequality is obviously satisfied. On the other hand, if $d(y) > 0$, then because $y \geq 0$ and $0 < r_1 < r_2$, we have $0 \leq y \leq y + r_1 d(y) < y + r_2 d(y)$, implying, because d is concave and increasing, that

$$\frac{d(y + r_1 d(y)) - d(y)}{r_1 d(y)} \geq \frac{d(y + r_2 d(y)) - d(y)}{r_2 d(y)},$$

which yields the desired inequality upon simplification. (We have used the fact that if f is concave increasing, then $[f(y + t) - f(y)]/t$ is decreasing in $t, t \geq 0$, for all y.) □

If at some time there are only two remaining items having ratings X and Y where

$$Y \underset{\text{LR}}{\leq} X,$$

then it follows from Lemmas 5.1 and 5.2 that issuing Y and then X stochastically maximizes the lifetime. That is, we have the following lemma.

Lemma 5.3 If at some time there are only two remaining items with ratings X and Y, where

$$Y \underset{\text{LR}}{\leq} X,$$

then issuing first Y and then X maximizes the probability that the remaining field life exceeds t for all t.

Before proving our main result we need the following additional lemma.

Lemma 5.4 If $d(t)$ is a nonnegative concave function, then $g_x(y) \equiv y + xd(y)$ is an increasing function of $y, y \geq 0$, for all $x \geq 0$.

Proof: The hypothesis implies that $d'(y) \geq 0$, and the result follows upon differentiation. □

We now show that the optimal policy is to issue items in increasing order of their likelihood ratios.

Proposition 5.5 Suppose that $d(t)$ is a nonnegative concave function. If the ratings $X_i, i = 1, \ldots, n$, are such that

$$X_1 \underset{\text{LR}}{\leq} X_2 \underset{\text{LR}}{\leq} \cdots \underset{\text{LR}}{\leq} X_n,$$

then the policy that first issues item 1, then item 2, then item 3, and so on, maximizes for all t the probability that the total field life exceeds t.

Proof: Because we have already proven the result for $n = 2$, assume it whenever there are $n - 1$ items to be issued. Now consider the n case, and suppose initially that we restrict attention to the $n!$ policies that specify the issuing orderings at time 0 (as opposed to policies whose decisions are allowed to depend on what has occurred up to then).

Consider any policy of this type, and let item j be the item that is issued last. Now, by Lemma 5.4 it follows that among all policies that issue item j last, the one yielding maximal field life is the one obtaining the largest possible field life from the other $n - 1$ items. Hence, by the induction hypothesis, of all policies that issue item j last, the one having the stochastically largest field life is the one that issues the first $n - 1$ items in increasing order of their likelihood ratios. Hence, if $j \neq n$, then the best policy that issues j last would issue item n next to last. However, if we let z denote the time when only items n and j remain, the problem at this point is one in which there are two items with a d function given by $d_z(t) = d(z + t)$. Because $d_z(t)$ is still concave and nonnegative, it follows from Lemma 5.3 that interchanging items j and n would lead to a stochastically larger field life. Hence, for any fixed-order policy, there is one that issues item n last and has a stochastically larger field life. But among all of those policies that issue n last, the one having the stochastically largest field life is, by the induction hypothesis and Lemma 5.4, the one given by the theorem.

We must now show this policy also stochastically maximizes lifetime, even when the decision is allowed to depend on the time at which it is made. (That is, for instance, after the first item has been issued we might allow the choice of the next item to depend on the lifetime of initial item.) Suppose we are allowed to let our decision depend on what has previously occurred at most k times. When $k = 1$ it follows from the preceding argument [because $d_y(t)$ is concave in t for all y] that it is optimal never to utilize this option. If this remains true when we are allowed $k - 1$ such opportunities, it follows by the same argument that it remains true when we are allowed k such opportunities (because by the induction hypothesis we would utilize such an opportunity at most once). Because the result is therefore true for all k, the result follows. □

6. A Stochastic Knapsack Model

A system must operate for t units of time. A certain component is essential for its operation and must be replaced, when it fails, with a new component. The class of spare components is grouped into n categories, with components of the ith category costing a positive amount C_i and functioning for an exponential length of time with rate λ_i. The problem of interest is, for a given t, to assign the initial component and subsequent replacements from among the n categories of spare components so as to minimize the expected cost of providing an operative component for t units of time.

We suppose that our surplus of spare parts contains an infinite number of each category. Letting $V(t)$ denote the minimal expected additional cost incurred when there are t time units to go and a failure has just occurred, then $V(t)$ satisfies the optimality equation

$$V(t) = \min_{i=1,\ldots,n} \left[C_i + \int_0^t V(t-x)\lambda_i e^{-\lambda_i x}\, dx \right], \qquad t > 0, \tag{6.1}$$

$$V(0) = 0.$$

Also, the policy that chooses, when t time units are remaining, a spare from a category whose number minimizes the right side of the optimality equation is an optimal policy.

Lemma 6.1 $V(t)$ is an increasing, continuous function of t for $t > 0$.

Proof: The increasing part follows from the definition of $V(t)$ because all costs are assumed nonnegative. To prove continuity, suppose that it is optimal to select a spare from category i whenever there are t units of time remaining. Then, by selecting this same category at time $t + \epsilon$, we see, using the lack of memory of the exponential and the monotonicity of V, and upon conditioning on whether or not the initial spare lasts for time ϵ, that

$$V(t) \le V(t + \epsilon) \le e^{-\lambda_i \epsilon} V(t) + (1 - e^{-\lambda_i \epsilon})[C_i + V(t + \epsilon)],$$

which establishes the result.   □

Let us suppose that all categories have distinct values of $\lambda_i C_i$, and suppose that they are numbered so that $\lambda_1 C_1 > \lambda_2 C_2 > \cdots > \lambda_n C_n$. (Problem 11 shows this is the only case we really need consider.) We are now ready for our main result, which states that the time interval $(0, \infty)$ can be broken up into n consecutive regions such that the optimal policy will call for a category i replacement whenever the remaining time is in the ith region.

Proposition 6.2 The optimal policy calls for the use of category-1 spares when the time remaining is small, then switches to category-2 spares as the time increases, then category-3 spares as the time further increases, etc., where, of course, the interval of use for some categories may be empty.

Proof: Consider any value $t > 0$. Suppose the assignment of category i when t units of time remain is uniquely optimal. Then, by the continuity of V and the optimality equation (6.1), there is an interval $(t, t + \epsilon), \epsilon > 0$, such that i is uniquely optimal at every point in $(t, t + \epsilon)$. Suppose several categories are optimal at t. Then the expressions within the brackets of (6.1) corresponding to each of the optimal categories are all equal to $V(t)$. If i is optimal, the derivative of the expression with respect to t corresponding to category i is

$$\frac{d}{dt}\left[C_i + \int_0^t V(t-x)\lambda_i e^{-\lambda_i x}\,dx \right] = \frac{d}{dt}\left[e^{-\lambda_i t} \int_0^t V(y)\lambda_i e^{\lambda_i y}\,dy \right]$$

$$= \lambda_i V(t) - \lambda_i \int_0^t V(y)\lambda_i e^{-\lambda_i(t-y)}\,dy$$

$$= \lambda_i V(t) - \lambda_i \int_0^t V(t-x)\lambda_i e^{-\lambda_i x}\,dx$$

$$= \lambda_i V(t) - \lambda_i[V(t) - C_i]$$

$$= \lambda_i C_i;$$

the derivative existing because $V(t)$ is continuous. It follows that, among those categories that are optimal at t, the category j with the smallest $\lambda_j C_j$ will be uniquely optimal over some interval $(t, t + \epsilon'), \epsilon' > 0$. Because at each change (as t increases) of optimal category, a category with a smaller λC becomes optimal, there can be at most n values of t for which a change in optimal category takes place. $\square$

Proposition 6.2 suggests the following algorithm for determining the optimal policy.

Algorithm 6.3   Let V_i denote the minimal expected cost function, and let π_i denote the optimal policy when only categories 1, 2, ..., i are available. For instance,

$$V_1(t) = C_1(1 + \lambda_1 t), \qquad 0 \leq t < \infty,$$

and π_1 is the policy that always replaces with a spare from category 1. From our previous structural results it follows that π_i will use category i whenever the time remaining is at least some finite critical value t_{i-1}. Now, at t_{i-1} it follows, by continuity, that it is optimal either to use category i and then proceed optimally or to just use π_{i-1}. Hence, if $t_{i-1} > 0$, then

$$V_{i-1}(t_{i-1}) = C_i + \int_0^{t_{i-1}} V_{i-1}(t_{i-1} - x)\lambda_i e^{-\lambda_i x}\, dx.$$

Furthermore, because it follows from the optimality equation that, for small values of t, π_i chooses the category with minimal value of C_k, we obtain that $t_{i-1} = 0$ if and only if $C_i = \min_{1 \leq k \leq i} C_k$. Hence, unless this is the case, t_{i-1} can be taken to be the smallest positive solution of the preceding equation. And, in addition, we have

$$V_i(t) = \begin{cases} V_{i-1}(t), & t \leq t_{i-1} \\ V_{i-1}(t_{i-1}) + \lambda_i C_i(t - t_{i-1}), & t \geq t_{i-1}, \end{cases} \tag{6.2}$$

and π_i uses category i whenever $t \geq t_{i-1}$ and follows π_{i-1} when $t \leq t_{i-1}$.

Thus we can recursively obtain π_i and $V_i(t)$ for $i = 1, \ldots, n$, where $V_n(t) = V(t)$, and π_n is the optimal policy.

7. A Sequential-Assignment Problem

Suppose there are n people available to perform n jobs, which arrive in sequential order. Associated with the jth job is a random variable X_j, $j = 1, \ldots, n$, and we suppose these random variables to be independent and identically distributed according to F. The ith person has a value p_i associated with him, and if this person is assigned to a job of value x, then an expected reward $p_i x$ is received. (One possibility is that the ith person will be able to perform a job successfully with probability p_i so, if assigned to a job of value x, will yield an expected reward of $p_i x$.) The problem is to assign the people to jobs as they arrive to maximize the expected total reward.

It follows by a result known as Hardy's theorem (stated later) that, if the values of the n jobs were known, then it would be optimal to assign the person with largest p to the job with largest value, the person with second highest p to the job of second highest value, and so on. Hence, it is intuitive that if the value of the initial job is large enough, then it is optimal to assign our best person (in the sense of highest p_i); if it is slightly smaller, then we should assign our next best person; and so on. What is not quite as intuitive, and what we shall show, is that the critical regions determining which person to assign do not depend on the p_i's.

We start with Hardy's theorem, the proof of which is left as an exercise.

Lemma 7.1 Hardy's Theorem   If $x_1 \leq x_2 \leq \cdots \leq x_n$ and $y_1 \leq y_2 \leq \cdots \leq y_n$, then

$$\max_{(i_1,\ldots,i_n)\in P} \sum_{j=1}^{n} x_{i_j} y_j = \sum_{j=1}^{n} x_j y_j,$$

where P is the set of all permutations of $(1, 2, \ldots, n)$.

Let $f(p_1, \ldots, p_n)$ denote the expected total reward earned under an optimal policy when the people to be assigned have values $p_1, \ldots, p_n$. Also, let $f(p_1, \ldots, p_n | x)$ denote this value conditional on the initial job having value x. We are now ready for our main result.

Proposition 7.2   For each $n \geq 1$, there exist numbers

$$-\infty = a_{0,n} \leq a_{1,n} \leq a_{2,n} \leq \cdots \leq a_{n,n} = +\infty$$

such that, whenever there are n stages to go and probabilities $p_1 \leq p_2 \leq \cdots \leq p_n$, then the optimal choice in the initial stage is to use p_i if the random variable X_i is contained in the interval $(a_{i-1,n}, a_{i,n}]$. The $a_{i,n}$ are independent of the p's. Furthermore, $a_{i,n}$ is the expected value, in an $(n-1)$-stage problem, of the quantity to which the ith smallest p is assigned under an optimal policy, and

$$f(p_1, p_2, \ldots, p_{n-1}) = \sum_{i=1}^{n-1} p_i a_{i,n} \qquad \text{for all } p_1 \leq p_2 \leq \cdots \leq p_{n-1}. \quad (7.1)$$

Proof: A proof by induction is employed. Suppose that there exist numbers $\{a_{j,m}\}_{j=1}^{m-1}$, $m = 1, 2, \ldots, n-1$, such that the optimal policy in an m-stage problem is initially to use the ith smallest p if the initial value is contained in the interval $(a_{i-1,m}, a_{i,m}]$, where $a_{0,m} = -\infty$ and $a_{m,m} = \infty$. Then, in the n-stage problem,

$$f(p_1, p_2, \ldots, p_n | x)$$
$$= \max_k [xp_k + f(p_1, p_2, \ldots, p_{k-1}, p_{k+1}, \ldots, p_n)]. \quad (7.2)$$

However, it follows by the induction hypothesis that the optimal policy for an $(n-1)$-stage problem is independent of the $n-1$ values of p. Hence, defining $a_{i,n}$, $i = 1, \ldots, n-1$, as the expected value (under the optimal policy) of the quantity to which the ith smallest p is assigned in the $(n-1)$-stage problem, the total expected reward of that problem is given by

$$f(\bar{p}_1, \bar{p}_2, \ldots, \bar{p}_{n-1}) = \sum_{i=1}^{n-1} \bar{p}_i a_{i,n} \quad (7.3)$$

for every $\bar{p}_1 \le \bar{p}_2 \le \cdots \le \bar{p}_{n-1}$ [the $\bar{p}_1$, $\bar{p}_2$, ..., $\bar{p}_{n-1}$ represent the remaining $n-1$ p's of the original n p's after the first is chosen in the n-stage problem]. Furthermore, because $a_{i,n}$ is independent of the p's and other policies are obtained by permuting the p's, any sum of the form $\sum_{i=1}^{n-1} \bar{p}_{j_i} a_{i,n}$ (where $j_1, j_2, \ldots, j_{n-1}$ is a permutation of the integers) can be obtained for the total expected reward of the $(n-1)$-stage problem. Hence, using Hardy's theorem, it follows that

$$a_{1,n} \le a_{2,n} \le \cdots \le a_{n-1,n} \quad (7.4)$$

because $f(\bar{p}_1, \bar{p}_2, \ldots, \bar{p}_{n-1})$ is a maximum.

Using the results of (7.3) and (7.4), Eq. (7.2) can now be expressed as

$$f(p_1, p_2, \ldots, p_n | x) = \max_k \left(xp_k + \sum_{i=1}^{k-1} p_i a_{i,n} + \sum_{i=k+1}^{n} p_i a_{i-1,n} \right). \quad (7.5)$$

Again, using Hardy's theorem (Lemma 7.1), we see that

$$f(p_1, p_2, \ldots, p_n | x) = xp_{k*} + \sum_{i=1}^{k*-1} p_i a_{i,n} + \sum_{i=k*+1}^{n} p_i a_{i-1,n},$$

where $k*$ is such that (with $a_{0,n} = -\infty$ and $a_{n,n} = +\infty$)

$$a_{k*-1,n} < x \le a_{k*,n}.$$

This result follows because the p's and a's are ordered so that, if x is greater than or equal to the $(k* - 1)$ smallest a, then the corresponding p (i.e., p_{k*}) must be greater than or equal to the $(k* - 1)$ smallest p. Hence, the first choice in an n-stage problem is to choose p_i if $x \in (a_{i-1,n}, a_{i,n}]$. Noting that the result is trivial for $n = 1$ completes the induction. Equation (7.1) follows immediately from Eq. (7.3) and the theorem is complete. $\square$

Whereas the previous proposition presents the structure of the optimal policy, it does not indicate how to obtain the $a_{i,n}$. However, as a simple consequence of Proposition 7.2, we have the following corollary.

Corollary 7.3

$$a_{i,n+1} = \int_{a_{i-1,n}}^{a_{i,n}} x \, dF(x) + a_{i-1,n} F(a_{i-1,n}) + a_{i,n} \bar{F}(a_{i,n}),$$

for $i = 1, \ldots, n$, where $-\infty \cdot 0 = \infty \cdot 0$ are defined as 0.

Proof: The result follows by recalling that $a_{i,n+1}$ is the expected value, in an n-stage problem, of the quantity to which the ith smallest p is assigned. The result then follows by conditioning on the initial x and recalling that p_i is used if and only if this value lies in the interval $(a_{i-1,n}, a_{i,n}]$. $\square$

Hence, starting with $a_{0,1} = -\infty$, the preceding can be used to compute recursively the values $a_{i,n+1}$. For instance,

$$a_{1,2} = \int_0^\infty x \, dF(x),$$

as was expected. (Why?)

Problems

1. For the model of Section 3, construct an example in which policy $(2, 1, i_k, \ldots, i_n)$ is better than policy $(1, 2, i_k, \ldots, i_n)$ even though $\mu_1 \le \mu_2$.

2. For the model of Section 4, show that

$$E(C_{1,2}) \le E(C_{2,1}) \Leftrightarrow \lambda_1 - \mu_1 \ge \lambda_2 - \mu_2.$$

3. Let $d(t)$, $t \ge 0$, be a nonnegative concave function. Show that this implies that $d(s) \le d(t)$ when $s \le t$.

4. Show that if f is a concave, increasing function, then

$$[f(x + t) - f(x)]/t$$

is decreasing in t, $t \geq 0$.

5. Does Proposition 5.1 remain valid when

$$Y \underset{st}{\leq} X \qquad \text{replaces} \qquad Y \underset{LR}{\leq} X.$$

Prove or give a counterexample. *Hint:* Try $h(y, x) = y + 2x$.

6. For the model of Section 5, show that
   (a) if d is increasing and convex, then, when $n = 2$, issuing item 2 and then 1 is optimal when $X_1 \leq_{LR} X_2$;
   (b) generalize the result in part (a) to the case of general n.

7. Suppose we have a stockpile of n items, the ith of which has a field life of X_i, where

$$X_1 \underset{LR}{\geq} X_2 \underset{LR}{\geq} \cdots \underset{LR}{\geq} X_n.$$

These items are to be released to the field in a sequential order. However, shocks occur according to a nonhomogeneous Poisson problem with intensity function $\lambda(t)$, and if a shock occurs while an item is in the field, then that item instantaneously dies and a new item must be placed in the field. The objective is to maximize the expected total field life.
   (a) If $\lambda(t)$ increases in t, determine the optimal policy.
   (b) If $\lambda(t)$ decreases in t, determine the optimal policy.

8. We have n missions to perform, the ith taking a random time X_i, where

$$X_1 \underset{LR}{\geq} X_2 \underset{LR}{\geq} \cdots \underset{LR}{\geq} X_n.$$

The missions are to be attempted sequentially. However, if a shock occurs while a mission is being attempted, then that mission must be unsuccessfully aborted and a new mission begun. Suppose shocks occur in accordance with a nonhomogeneous Poisson process having intensity $\lambda(t)$. If $\lambda(t)$ increases in t, what ordering maximizes the expected number of missions successfully performed? What if $\lambda(t)$ decreases?

9. Consider a stockpile of two items having initial ages x_1 and x_2.

The items are issued sequentially, and the objective is to maximize the expected total field life, when it is supposed that when an item of age x is put in the field, its field life will be an exponential random variable with mean x. Determine the optimal policy.

10. For the stochastic knapsack model of Section 6, show that the optimal value function V is piecewise linear and concave.

11. For the model of Section 6
    (a) verify the expression given for $V_1(t)$;
    (b) compute $V_2(t)$;
    (c) explain Eq. (6.2);
    (d) show that if $C_j \geq C_i$ and $\lambda_j > \lambda_i$, then category j is never used in an optimal policy;
    (e) Show that if $\lambda_i C_i \leq \lambda_j C_j$ and $C_i < C_j$, then category j will never be used in an optimal policy.

12. Prove Hardy's theorem.

13. Compute the optimal policy for the model of Section 7 when $n = 4$ and $F(x) = x, 0 < x \leq 1$.

Notes and References

Section 2 is taken from Derman *et al.* [3]. The proof of Theorem 3.2 is taken from Pinedo and Weiss [5]. The generalization to an arbitrary number of servers is more involved and has been given in a few papers, see, for instance, Weiss and Pinedo [8] or Weber [6]. The stronger result—that the policy given in Theorem 3.2 stochastically minimizes the makespan and stochastically maximizes the time until one of the processors becomes idle—can be found in Weber [6]. Theorem 3.3 appears to be new. For the model of Section 4, we followed the approach given by Weiss [7], which presents a nice survey of multiserver scheduling models. Another survey of such models is given by Pinedo and Schrage [4]. Section 5 is taken from Brown and Solomon [1]. The stochastic knapsack model of Section 6 is from Derman *et al.* [3]; and the sequential-assignment model of Section 7 from Derman *et al.* [2].

1. Brown, M. and Solomon, H., "Optimal issuing policies under stochastic field lives," *J. Appl. Prob.* **10**, 761–768, 1973.

2. Derman, C., Lieberman, G. J., and Ross, S. M., "A sequential stochastic assignment model," *Management Sci.* **18** (7), 349–355, 1972.

3. Derman, C., Lieberman, G. J., and Ross, S. M., "A renewal decision problem," *Management Sci.* **24** (5), 554–562, 1978.

4. Pinedo, M. and Schrage, L., "Stochastic shop scheduling: a survey," in *Deterministic and Stochastic Scheduling*, eds. M. Dempster, J. K. Lenstra, and A. Rinooy-Kan. D. Reidel, Dordrecht, Holland, 1982.

5. Pinedo, M. and Weiss, G., "Scheduling of stochastic tasks on two parallel processors," *Naval Res. Logist. Quart.* **26,** 527–535, 1971.

6. Weber, R. R., "Scheduling jobs with stochastic processing requirements on parallel machines to minimize makespan or flowtime," *J. Appl. Probab.* **19** (1), 167–182, 1982.

7. Weiss, G., "Multiserver stochastic scheduling," in *Deterministic and Stochastic Scheduling*, eds. M. Dempster, J. K. Lenstra, and A. Rinooy-Kan. D. Reidel, Dordrecht, Holland, 1982.

8. Weiss, G. and Pinedo, M., "Scheduling tasks with exponential service times on non-identical processors to minimize various cost functions," *J. Appl. Probab.* **17,** 187–202, 1980.

VII

Bandit Processes

1. Introduction

Suppose we have n independent projects on which we can work. Each project is, at any time, in some state. If the state of a project is i and we decide to work on that project, then we receive an expected reward $R(i)$ and the next state of that project becomes j with probability P_{ij}. Moreover, the remaining $n - 1$ projects, which are not worked on, do not change state. In addition, we allow ourselves the option of retiring, which earns us M and ends the problem.

In this chapter we show how to determine the optimal policy by first solving the single-project case $n = 1$.

2. Single-Project Bandit Processes

Consider a project that at any time point is in some state. After observing the state, the decision maker must decide whether to operate the project or retire. If the state is i and the decision to continue (i.e., operate) is chosen, then a bounded reward $R(i)$ is earned and the process then goes to state k with probability P_{ik}. If the decision to retire (or stop) is made, then a terminal reward M is earned. Let $V(i:M)$ denote the maximal expected α-discounted return when the initial state is i, where α, $0 < \alpha < 1$, is the discount factor and where we use the notation $V(i:M)$ to indicate the dependence of the optimal value function on the terminal reward M. Then V satisfies the optimality equation

$$V(i:M) = \max\left[M, R(i) + \alpha \sum_k P_{ik} V(k:M)\right].$$

Our objective in this section is to prove that, if retirement is optimal in a given state when the termination reward is M, then it is also optimal whenever the termination reward is greater than M.

We start with the following lemma.

Lemma 2.1   For fixed i, $V(i:M) - M$ is decreasing in M.

Proof: Let $V_n(i:M)$ denote the maximal expected return when one is only allowed to continue for at most n stages before retiring. Then

$$V_0(i:M) = M,$$

$$V_n(i:M) = \max\left[M, R(i) + \alpha \sum_k P_{ik} V_{n-1}(k:M)\right].$$

We now show by induction on n that $V_n(i:M) - M$ is decreasing in M. It is obvious for $n = 0$, so assume it for $n - 1$. Now,

$$V_n(i:M) - M = \max\{0, R(i) + \alpha \sum_k P_{ik}\left[V_{n-1}(k:M) - M\right] - (1 - \alpha)M\},$$

and the result follows from the induction hypothesis. $\square$

It follows from Lemma 2.1 that, if for fixed initial state i it is optimal to retire when the termination reward is M, then it would also be optimal to retire if this reward were greater than M. This follows because, for $\overline{M} \geq M$, we have

$$0 \leq V(i:\overline{M}) - \overline{M} \qquad \text{by the optimality equation}$$

$$\leq V(i:M) - M \qquad \text{by Lemma   2.1}$$

$$= 0 \qquad \text{because retirement is optimal at } M,$$

so $V(i:\overline{M}) = \overline{M}$.

Let $M(i)$ denote the smallest value of M at which retirement at i is optimal. That is,

$$M(i) = \inf\left[M: V(i:M) = M\right].$$

In fact, because $V(i:M)$ is continuous in M (immediate by induction), we can write

$$M(i) = \min\left[M: V(i:M) = M\right].$$

Hence, using this, the optimal policy can be described as follows.

Theorem 2.2 When in state i, it is optimal to retire if $M \geq M(i)$ and to continue if $M \leq M(i)$. Hence, $M(i)$ represents the indifference value for state i. It is the retirement termination value that makes one indifferent between retiring or continuing when in state i.

3. Multiproject Bandit Processes

Let us now suppose that, rather than a single project, there are n identical projects available, each of which is in some state. (The restriction to identical projects is made for notational convenience and, as we will see later, actually introduces no loss of generality.) After observing the state of each project, a decision must be made to operate on exactly one of the projects or to retire. If a project whose state is i is operated upon, then (i) a reward $R(i)$ is earned, (ii) that project then enters state k with probability P_{ik}, and (iii) the states of the other projects remain the same. If the decision to retire is made, then the process ends and a termination reward M is earned.

If we let $V(i:M)$ denote the maximal expected α-discounted return when the initial project states are $\mathbf{i} = (i_1, \ldots, i_j, \ldots, i_n)$, then V satisfies

$$V(\mathbf{i}:M) = \max[M, \max_j O^j(V(\mathbf{i}:M)],$$

where $O^j(V(\mathbf{i}:M)$, the maximal expected return if project j is operated on, is given by

$$O^j(V(\mathbf{i}:M)) = R(i_j) + \alpha \sum_k V(i_1, \ldots, i_{j-1}, k, i_{j+1}, \ldots, i_n:M)P_{i_jk}.$$

The major difficulty in numerically solving for V (say, by use of successive approximations) is that the number of states becomes large very quickly. For instance, if each individual project has K possible states, then the multiproject problem would have K^n states. Hence, to solve a multiproject bandit problem directly is usually not computationally feasible. However, it turns out that such a problem can be decomposed into its n constituent single-project bandit problems. In fact, if we let $M(i)$ denote the indifference value for state i when only one project is available, we shall show that the optimal policy for the multiproject bandit problem is as follows: If the state is $\mathbf{i} = (i_1, i_2, \ldots, i_n)$, it is optimal to

retire if $M(i_j) \leq M$ for   all $j = 1, \ldots, n,$

operate on j if $\max_{k} M(i_k) = M(i_j) > M.$

To prove that this is indeed optimal, we start by studying the properties of $V(\mathbf{i}:M)$ as a function of M. We then speculate about the conditions under which it is optimal to retire and show that the results of our speculation imply a particular expression for $V(\mathbf{i}:M)$. We then prove the correctness of this expression by showing that it satisfies the optimality equation, and in doing so we also prove the optimality of the preceding policy.

Lemma 3.1   For fixed $\mathbf{i}$, $V(\mathbf{i}:M)$ is an increasing, convex function of M.

Proof: The increasing part is obvious because increasing the terminal reward cannot decrease the maximal expected return. To prove convexity, consider any stationary policy π, and for a given initial state $\mathbf{i}$, let T_π denote the retirement time under π. That is, T_π is the number of stages that projects are operated before retirement. Then,

$$V_\pi(\mathbf{i}:M) = E_\pi(\text{discounted return prior to time } T_\pi + M\alpha^{T_\pi}).$$

Hence,

$$V(\mathbf{i}:M) = \max\big[E_\pi(\text{discounted return prior to } T_\pi) + ME(\alpha^{T_\pi})\big], \quad (3.1)$$

and because $V(\mathbf{i}:M)$ is the maximum of terms, each of which is linear in M, it is thus convex in M (see Problem 1).   □

Lemma 3.2   For fixed initial state $\mathbf{i}$, let T_M denote the optimal retirement time under the policy for terminal reward M. Then, for all M for which $\partial V(\mathbf{i}:M)/\partial M$ exists,

$$\frac{\partial}{\partial M} V(\mathbf{i}:M) = E(\alpha^{T_M} \,|\, X_0 = \mathbf{i}).$$

Proof: Fix M and the initial state $\mathbf{i}$. Let $\bar{\pi}$ denote an optimal policy and let T_M denote the number of stages of project operation before retirement. If we employ $\bar{\pi}$ for a problem having terminal reward $M + \epsilon, \epsilon > 0$, we receive

$$E_\pi\big[\text{return prior to } T_M\big] + (M + \epsilon)E(\alpha^{T_M}).$$

Hence, using (3.1) and the fact that the optimal return when the terminal reward is $M + \epsilon$ is at least that given by the preceding expression, we obtain

$$V(\mathbf{i}:M + \epsilon) \geq V(\mathbf{i}:M) + \epsilon E(\alpha^{T_M}). \tag{3.2}$$

Similarly, using the M-optimal policy for a problem having terminal retirement reward $M - \epsilon$ *gives*

$$V(\mathbf{i}:M - \epsilon) \geq V(\mathbf{i}:M) - \epsilon E[\alpha^{T_M}]. \tag{3.3}$$

Hence, from (3.2) and (3.3) we see that

$$\frac{V(\mathbf{i}:M + \epsilon) - V(\mathbf{i}:M)}{\epsilon} \geq E[\alpha^{T_M}]$$

and

$$\frac{V(\mathbf{i}:M) - V(\mathbf{i}:M - \epsilon)}{\epsilon} \leq E[\alpha^{T_M}].$$

Hence, the result follows upon letting $\epsilon \to 0$. $\square$

Technical Remark: Because $V(\mathbf{i}:M)$ is a convex function of M, it follows that $\partial V(\mathbf{i}:M)/\partial M$ exists at almost all values of M.

If we let $V(i:M)$ denote the optimal value function when only a single project is available and let T denote the optimal retirement time, then it follows from Lemma 3.2 that

$$\frac{\partial}{\partial M} V(i:M) = E(\alpha^T \mid X_0 = i).$$

Now let $M(i)$ be the indifference value when only a single project is available and its state is i. Consider the multiproject case, suppose the state is $\mathbf{i} = (i_1, \ldots, i_j, \ldots, i_n)$, and let us speculate about whether or not we would ever again operate project j. If $M(i_j) > M$, then, because it would not be optimal to retire even if projedt j were the only project available, it is clear that we would never retire before operating project j. On the other hand, what if $M(i_j) \leq M$? Would we ever want to operate project j under this circumstance? Whereas it is not obvious that we would never operate project j when $M(i_j) \leq M$, it does seem somewhat intuitive, so let us accept this as a working hypothesis and see where it leads. That is, let us suppose that once a project reaches a state under which retirement would be optimal if it were the sole project available, then the optimal policy never again operates that project. From this it follows that the optimal policy would retire in state $\mathbf{i}$ when $M(i_j) \leq M$,

$j = 1, \ldots, n$. In fact, because it cannot be optimal to retire in **i** when $M(i_j) > M$ for some $j = 1, \ldots, n$ (because it is not optimal even if only project j were available), our speculations lead to the hypothesis that

(i) project j would never be operated if the state **i** is such that $M(i_j) \leq M$ and

(ii) retirement should occur if and only if $M(i_j) \leq M$ for all $j = 1, \ldots, n$.

For a given initial state $\mathbf{i} = (i_1, \ldots, i_j, \ldots, i_n)$, let T^j denote the optimal retirement time when only project j is available $j = 1, \ldots, n$. That is, T^j is the time project j has to be operated upon, when its initial state is i_j, until it reaches a state i such that $M(i) \leq M$. Also, let T denote the optimal retirement time for the multiproject case. Because the changes of state of individual projects are in no way affected by what occurs in other projects, it follows that, under our working hypothesis,

$$T = \sum_{j=1}^{n} T^j;$$

and, in addition, $T^j, j = 1, \ldots, n$ are independent random variables. Hence,

$$E(\alpha^T) = E(\alpha^{\Sigma_j T^j})$$

$$= \prod_{j=1}^{n} E(\alpha^{T^j}),$$

and from Lemma 3.2 we obtain

$$\frac{\partial}{\partial M} V(\mathbf{i}:M) = \prod_{j=1}^{n} \frac{\partial}{\partial M} V(i_j:M). \tag{3.4}$$

Now, let $(1 - \alpha)C$ be a bound on all one-period rewards $R(i)$. For $M \geq C$, note that any policy that does not call for immediate retirement has an expected return bounded by

$$\sup_{n>0} \left[(1 - \alpha)C + \alpha(1 - \alpha)C + \cdots + \alpha^{n-1}(1 - \alpha)C + \alpha^n M \right]$$

$$= (1 - \alpha)C \frac{(1 - \alpha^n)}{1 - \alpha} + \alpha^n M$$

$$\leq M.$$

Hence, for $M \geq C$ we see that immediate retirement is optimal. That is,

$$V(\mathbf{i}:M) = M \qquad \text{for} \quad M \geq C.$$

Integrating (3.4) yields

$$\int_M^C \frac{\partial}{\partial m} V(\mathbf{i}:m)\, dm = \int_M^C \prod_{j=1}^n \frac{\partial}{\partial m} V(i_j:m)\, dm$$

or

$$C - V(\mathbf{i}:M) = \int_M^C \prod_{j=1}^n \frac{\partial}{\partial m} V(i_j:m)\, dm.$$

Hence, our working hypothesis that a project is never again operated once it is in a state such that it would be retired if it were the only project available leads to the conclusion that V is given by

$$V(\mathbf{i}:M) = C - \int_M^C \prod_{j=1}^n \frac{\partial}{\partial m} V(i_j:m)\, dm. \qquad (3.5)$$

We now prove that (3.5) is indeed valid by showing that it satisfies the optimality equation, and in doing so, we shall obtain the structure of the optimal policy.

Theorem 3.3

$$V(\mathbf{i}:M) = C - \int_M^C \prod_{k=1}^n \frac{\partial}{\partial m} V(i_k:m)\, dm, \qquad M \leq C,$$

and, when in state $\mathbf{i}$, the optimal policy retires if $M(i_j) \leq M$ for all $j = 1, \ldots, n$ and operates project j if $M(i_j) = \max_k M(i_k) > M$.

Proof: Let

$$\hat{V}(\mathbf{i}:M) = C - \int_M^C \prod_{k=1}^n \frac{\partial}{\partial m} V(i_k:m)\, dm.$$

We shall show that $\hat{V}$ satisfies the optimality equation

$$\hat{V}(\mathbf{i}:M) = \max[M, \max_k O^k(\hat{V}(\mathbf{i}:M))],$$

which, by uniqueness (Proposition 2.3 of Chapter 2), shows that $\hat{V} = V$.

Fix $j, j = 1, \ldots, n$, and define

$$P_j(\mathbf{i}:M) = \prod_{k \neq j} \frac{\partial}{\partial M} V(i_k:M).$$

Because, from Lemma 3.1, $V(i_k:M)$ is an increasing and convex function of M, it follows that for fixed $\mathbf{i}$ and j, $P_j(\mathbf{i}:M)$ is a nonnegative, increasing (from convexity) function of M. Because $\partial V(i_k:M)/\partial M = 1$ for $M \geq C$, it follows that $P_j(\mathbf{i}:M) = 1$ for $M \geq C$.

Because $\hat{V}$ can be written as

$$\hat{V}(\mathbf{i}:M) = C - \int_M^C \frac{\partial}{\partial m} V(i_j:m) P_j(\mathbf{i}:m)\, dm,$$

we see, by integration by parts, that

$$\hat{V}(\mathbf{i}:M) = C - P_j(\mathbf{i}:m)V(i_j:m)\Big|_{m=M}^{m=C} + \int_M^C V(i_j:m)\, dP_j(\mathbf{i}:m)$$

or

$$\hat{V}(\mathbf{i}:M) = P_j(\mathbf{i}:M)V(i_j:M) + \int_M^C V(i_j:m)\, dP_j(\mathbf{i}:m), \qquad (3.6)$$

where we have used that $P_j(\mathbf{i}:C) = 1$, $V(i_j:C) = C$, and the final integral in the preceding expression is over m. Similarly, we have

$$O^j(\hat{V}(\mathbf{i}:M)) = R(i_j) + \alpha \sum_k P_{i_j,k} \hat{V}(i_1,\ldots,i_{j-1},k,i_{j+1},\ldots,i_n:M)$$

$$= R(i_j) + \alpha \sum_k P_{i_j,k}\big[P_j(\mathbf{i}:M)V(k:M)$$

$$+ \int_M^C V(k:m)\, dP_j(\mathbf{i}:m)\big], \qquad (3.7)$$

where we have used (3.6) and the fact that

$$P_j(i_1,\ldots,i_{j-1},k,i_{j+1},\ldots,i_n:M) = P_j(\mathbf{i}:M).$$

Hence, from (3.6) and (3.7), we see that

$$\hat{V}(\mathbf{i}:M) - O^j(\hat{V}(\mathbf{i}:M)) = P_j(\mathbf{i}:M)\big[V(i_j:M) - R(i_j) - \alpha \sum_k P_{i_j,k}V(k:M)\big]$$

$$+ \int_M^C \big[V(i_j:m) - \alpha \sum_k P_{i_j,k}V(k:m) \qquad (3.8)$$

$$- R(i_j)\big]\, dP_j(\mathbf{i}:m).$$

Now,

$$V(i_j:M) - R(i_j) - \alpha \sum_k P_{i_j,k}V(k:M) \geq 0, \qquad (3.9)$$

which shows that

$$\hat{V}(\mathbf{i}:M) \geq O^j(\hat{V}(\mathbf{i}:M)). \tag{3.10}$$

Let us see under what conditions equality occurs in (3.10). First, note that (3.9) holds with equality if nonretirement is optimal when only project j is available, that is, when $M \leq M(i_j)$. In this case, we have from (3.8) that

$$\hat{V}(\mathbf{i}:M) - O^j(\hat{V}(\mathbf{i}:M))$$

$$= \int_{M(i_j)}^{C} \Big[V(i_j:m) - R(i_j)$$

$$- \alpha \sum_k P_{i_j,k} V(k:m) \Big] dP_j(\mathbf{i}:m), \qquad \text{if} \quad M \leq M(i_j). \tag{3.11}$$

Now

$$P_j(\mathbf{i}:m) = \prod_{k \neq j} \frac{\partial}{\partial m} V(i_k:m) = 1 \qquad \text{for} \quad m \geq \max_{k \neq j} M(i_k),$$

so,

$$dP_j(\mathbf{i}:m) = 0 \qquad \text{for} \quad m \geq \max_{k \neq j} M(i_k).$$

Hence, using this, we see from (3.11) that

$$\hat{V}(\mathbf{i}:M) = O^j(\hat{V}(\mathbf{i}:M)) \qquad \text{if} \quad M \leq M(i_j) = \max_k M(i_k). \tag{3.12}$$

In addition, using (3.6) and the monotonicity of $V(k:m)$ in m, we obtain

$$\hat{V}(\mathbf{i}:M) \geq P_j(\mathbf{i}:M)V(i_j:M) + V(i_j:M)\int_M^C dP_j(\mathbf{i}:m)$$

$$= V(i_j:M) \tag{3.13}$$

$$\geq M.$$

In fact, inequality (3.13) holds with equality if $M \geq \max_k M(i_k)$. This is so because in this case $V(i_k:m) = m$ for $m \geq M$, $k = 1, \ldots, n$, and thus $P_j(\mathbf{i}:m) = 1$, implying that $dP_j(\mathbf{i}:m) = 0$ for $m \geq M$. Hence from (3.6) we see that

$$\hat{V}(\mathbf{i}:M) = M \qquad \text{if} \quad M \geq \max_k M(i_k). \tag{3.14}$$

Therefore, from (3.10), (3.12), (3.13), and (3.14), we see that $\hat{V}$ satisfies the optimality equation

$$\hat{V}(\mathbf{i}:M) = \max[M, \max_j O^j(\hat{V}(\mathbf{i}:M))],$$

so $\hat{V} = V$. That the optimal policy is as stated now follows from (3.12) and (3.14). $\square$

The preceding theorem shows that the optimal policy in the multi-project case can be determined by an analysis of the n single-project problems, with the optimal decision in state $\mathbf{i}$ being to operate on that project j having the largest indifference value $M(i_j)$ for its state i_j if this value is greater than M and to retire otherwise.

Another interpretation of the indifference value, which is quite illuminating, can be obtained as follows: Consider a single-project bandit process having initial state i. Because when $M = M(i)$ the optimal policy is indifferent between retiring and continuing, we have that for any positive random retirement time T, $M(i) \geq E$ (discounted return prior to T) + $M(i)E(\alpha^T)$, with equality for the optimal continuation policy. (This inequality results from the fact that retiring in i is optimal because $M = M(i)$, and the equality results because the best continuation rule is also optimal.) Hence,

$$M(i) \geq \frac{E[\text{discounted return prior to } T]}{[1 - E(\alpha^T)]},$$

with equality for the optimal continuation policy. Thus we may write

$$M(i) = \max_{T>0} \frac{E[\text{discounted return prior to } T]}{1 - E(\alpha^T)},$$

so

$$(1 - \alpha)M(i) = \max_{T>0} \frac{E(\text{discounted return prior to } T)}{E[1 - \alpha^T)/(1 - \alpha)]}$$

$$= \max_{T>0} \frac{E(\text{discounted return prior to } T)}{E(1 + \alpha \cdots + \alpha^{T-1})}$$

$$= \max_{T>0} \frac{E(\text{discounted return prior to } T)}{E(\text{discounted time prior to } T)},$$

where the expectations are conditional on the initial state being i.

Hence, another way of describing the optimal policy in the multi-project case is as follows: For each individual project look for the retirement time T whose ratio of expected discounted return prior to T to the expected discounted time prior to T is maximal. Then consider the largest such ratio obtained; if it is greater than $(1 - \alpha)M$, then operate on the corresponding project, if not, retire.

We shall now establish that the assumption of identical projects is without any loss of generality. Suppose the projects were not identical but rather, when project j is operated in state i, a reward $R^j(i)$ is earned and the next state of that project is k with probability P^j_{ik}. Such a situation can be treated within the identical-project case by expanding the state space to consist of n blocks, one for each project, such that transitions from one block to another never occur. That is, let the state-space be $\{(j,i),\ j = 1, \ldots, n\}$, and take $P_{(j,i)(j,k)} = P^j_{ik}$, $R(j,i) = R^j(i)$. Therefore, the jth bandit process can be regarded as a bandit process of the preceding (extended state space) type, with its initial state being one of the states (j,i). Hence, all of our results extend to the case of nonidentical processes.

In specific examples having nonidentical projects, we shall utilize the notation $M^j(i)$ to be the indifference value for state i when only project j is available.

EXAMPLE 3.1a   Suppose an individual has n jobs to perform. He will receive $R(j)$, $R(j) > 0$, when job j is completed. The amount of time required for completion of job j, call it X_j, is a random variable with

$$P(X_j = k) = \beta^j(k), \qquad k = 1, 2, \ldots.$$

At each time point the individual must decide upon which job to work. He is allowed to switch jobs at any time and, when he returns to a job, we suppose that work picks up at the point it left off. For instance, suppose $P(X_j = 5) = P(X_j = 20) = \frac{1}{2}$, and suppose 5 time units are spent working on job j without its completion. If work then begins on another job, then job j will still need an additional 15 units of work when the individual returns to it.

This is a multiproject bandit problem in which operating project j corresponds to working on job j. Project j will be in state i if i time units of work have been performed on j and it has not yet been finished. The retirement return is $M = 0$ (and thus retirement will never occur).

Hence, we have

$$(1 - \alpha)M^j(i) = \max_{T > 0} \frac{E(\text{discounted return prior to } T \,|\, X_0 = i)}{E(\text{discounted time prior to } T \,|\, X_0 = i)}. \quad (3.15)$$

Also, because the only randomness in this problem is whether or not a job is completed, it follows that a stopping time just specifies a deterministic number k with the interpretation that the project is to be worked on for either an additional time k or until completion, whichever occurs first. Thus from (3.15) we have

$$(1 - \alpha)M^j(i) = \sup_{k > 0} \frac{\sum_{m=1}^{k} P(X_j = i + m \,|\, X_j > i)\alpha^{m-1} R(j)}{\sum_{m=1}^{k} P(X_j = i + m \,|\, X_j > i)(1 + \alpha + \cdots + \alpha^{m-1})}$$
$$+ P(X_j > i + k \,|\, X_j > i)(1 + \alpha + \cdots + \alpha^{k-1})$$

$$= \sup_{k > 0} \frac{\sum_{m=1}^{k} P(X_j = i + m \,|\, X_j > i)\alpha^{m-1} R(j)}{\sum_{m=1}^{k} P(X_j \geq i + m \,|\, X_j > i)\alpha^{m-1}}$$

$$= \sup_{k > 0} \frac{\sum_{m=1}^{k} P(X_j = i + m)\alpha^{m-1} R(j)}{\sum_{m=1}^{k} P(X_j \geq i + m)\alpha^{m-1}},$$

where we have made use of the identity

$$\sum_{m=1}^{k} P(Z \geq m)\alpha^{m-1} = \sum_{m=1}^{k} P(Z = m)(1 + \alpha + \cdots + \alpha^{m-1})$$
$$+ P(Z > k)(1 + \alpha + \cdots + \alpha^{k-1}).$$

The optimal policy when in state $\mathbf{i} = (i_1, \ldots, i_n)$ is, of course, to work on the job j not yet completed having the largest value of $M^j(i_j)$.

Special Case (i): Increasing failure rate.   In the special case in which X_j has a discrete increasing failure rate in the sense that $\lambda_j(l) \equiv P(X_j = l)/P(X_j \geq l)$ increases in l, it can be shown that the supremum is attained for $k = \infty$, and so

$$(1 - \alpha)M^j(i) = \frac{\sum_{m=1}^{\infty} P(X_j = i + m)\alpha^{m-1} R(j)}{\sum_{m=1}^{\infty} P(X_j \geq i + m)\alpha^{m-1}}$$

Indeed, it can also be shown (Problem 4) that in this case $M^j(i)$ increases in i. Hence, assuming that all job-completion times have increasing failure rate, it follows that if initially no work has been performed on any of the jobs, then the jobs should be scheduled in decreasing order of

$$(1 - \alpha)M^j(0) = \frac{\sum_{m=1}^{\infty} P(X_j = m)\alpha^{m-1}R(j)}{\sum_{m=1}^{\infty} P(X_j \geq m)\alpha^{m-1}}$$

or, equivalently, using (3.17), in decreasing order of

$$M^j(0) = \frac{R(j)E(\alpha^{X_j})}{\alpha E(1 - \alpha^{X_j})}.$$

In addition, once started, a job should be continued to completion (because $M^j(i + 1) \geq M^j(i)$, it follows that if $M^j(i_j)$ is the maximal present indifference value, then $M^j(i_j + 1)$ will be maximal in the next period).

Special Case (ii): Decreasing failure rate.   If X_j has a discrete decreasing failure rate function in that $\lambda_j(l) \equiv P(X_j = l | X_j \geq l)$ decreases in l, it follows (see Problem 5) that the supremum in (3.16) occurs at $k = 1$, so

$$(1 - \alpha)M^j(i) = \lambda_j(i + 1)R(j).$$

4. An Extension and a Nonextension

4.1. Restricted States

Suppose that when a certain project enters a given set of states we are required to operate it until it leaves that set. For instance, suppose it is not allowed to leave project 1 in any of the states in S. Then if this project is operated and its state enters S, we are required to continue operating it at least until its state leaves S. It is easy to verify that our results go through as before (see Problem 6) and that Theorem 3.3 remains valid. [Of course, the indifference function $M(i)$ reflects the fact that we are restricted from leaving the project in certain states.]

EXAMPLE 4.1A   Let us reconsider Example 3.1a, in which there are n jobs to be processed sequentially. The ith job takes a time X_i and earns upon completion, R_i. However, suppose now that a job is not allowed to be preempted, that is, once a job is begun it must be continued until completion. What is the optimal ordering in this case?

We can regard this as an example of a bandit problem with restricted states. The state of any project (i.e., job) is x, the amount of time already spent working on that job. If no preemptions are allowed, then all states except 0 and ∞ (where ∞ signifies that the job is completed) are restricted. Because $M_i \equiv M_i(0)$, the indifference value for the ith job is such that when

$$M_i = E[\alpha^{X_i}(R_i + M_i)],$$

we see that

$$M_i = \frac{R_i E(\alpha^{X_i})}{1 - E(\alpha^{X_i})},$$

and the optimal policy is to process the jobs in decreasing order of their M_i.

4.2. A Nonextension

In many applications it is natural to assume that several projects can be simultaneously operated upon. For instance, if projects correspond to jobs that must be performed, then our analysis so far presumes only a single worker (or machine), so the jobs must be performed sequentially. However, it is often the case that many workers (or machines) are available, so jobs can be simultaneously performed. At first glance one might hope that it is optimal to operate on those projects having the largest indifference value. That this need not be the case is shown by the following counterexample.

EXAMPLE 4.1b   Suppose there are three jobs to be performed, each having a fixed deterministic processing time and a return upon completion that is given by the following tabulation.

Job	Processing time	Worth when processed
1	2	1
2	1	1
3	2	21/10

Letting the discount factor α equal $\frac{1}{2}$ gives that M^i, the indifference value for job i, is such that

$$M^1 = \tfrac{1}{2} + (\tfrac{1}{2})^2 M^1 \qquad \text{or} \quad M^1 = \tfrac{2}{3} = M^2,$$

$$M^3 = (\tfrac{1}{2})^2 (\tfrac{21}{10}) + (\tfrac{1}{2})^3 M^3 \qquad \text{or} \quad M^3 = \tfrac{3}{5}.$$

Hence, $M^1 = M^2 > M^3$.

Now suppose that there are two machines available for processing jobs. If the order 1, 2, 3 is used, then the expected return is

$$\tfrac{1}{2} + \tfrac{1}{2} + (\tfrac{1}{2})^3 (\tfrac{21}{10}) = \tfrac{101}{80};$$

whereas if the order 1, 3, 2 is used then the expected return is

$$\tfrac{1}{2} + (\tfrac{1}{2})^2 (\tfrac{21}{10}) + (\tfrac{1}{2})^2 = \tfrac{102}{80}.$$

Hence, with two machines it is no longer optimal to order according to the M^i.

5. Generalizations of the Classical Bandit Problem

In the classical one-arm bandit problem, on each play of the game one either wins or loses 1 unit. It is supposed that the win probability p is unknown and has been chosen from a prior distribution. Rewards are discounted according to α, $0 < \alpha < 1$, and in addition, there is available at each time point an option to quit and receive a terminal reward M. One application of this model is to a slot machine (commonly called a one-arm bandit) having an unknown win probability. A second and more serious application is to medical trials of a new drug whose cure probability has not yet been determined. In this latter case, quitting may mean reverting to a known drug whose cure rate has been established.

Let us generalize this model to a situation in which the amount earned on each play is a random variable whose density function $f(x|\theta)$ depend on the value of a parameter θ, which is itself regarded as a random value chosen from a prior distribution having density $g(\theta)$. We also suppose a retirement option that earns us M, and we discount rewards according to α.

We shall suppose that the conditional density $f(x|\theta)$ increases in likelihood ratio as θ increases or, equivalently,

$$\frac{f(x|\theta_1)}{f(x|\theta_2)} \quad \text{increases in } x \qquad \text{for} \quad \theta_1 > \theta_2. \tag{5.1}$$

EXAMPLE 5.1a (i)   In the classical bandit problem,

$$f(1|\theta) = \theta, \qquad f(-1|\theta) = 1 - \theta,$$

so

$$\frac{f(1|\theta_1)}{f(1|\theta_2)} \geq \frac{f(-1|\theta_1)}{f(-1|\theta_2)} \Leftrightarrow \frac{\theta_1}{\theta_2} \geq \frac{1-\theta_1}{1-\theta_2} \Leftrightarrow \theta_1 \geq \theta_2,$$

verifying (5.1)

(ii)   If the rewards are normally distributed with unknown mean θ and known variance, say, equal to σ^2, then

$$f(x|\theta) = \frac{1}{\sqrt{2\pi}\sigma} \exp\left[\frac{-(x-\theta)^2}{2\sigma^2}\right].$$

Hence, for $\theta_1 > \theta_2$,

$$\frac{f(x|\theta_1)}{f(x|\theta_2)} = \exp\left\{\frac{1}{2\sigma^2}\left[(x-\theta_2)^2 - (x-\theta_1)^2\right]\right\}$$

$$= \exp\left\{\frac{1}{2\sigma^2}\left[2(\theta_1 - \theta_2)x + \theta_2^2 - \theta_1^2\right]\right\}$$

which is seen to increase in x, so (5.1) is satisfied.

Let $g(\theta)$ denote a prior density on θ, and suppose that, given θ, X is distributed according to the density $f(x|\theta)$. If we let $f_g(x)$ denote the unconditional density of X, then

$$f_g(x) = \int f(x|\theta)g(\theta)\,d\theta.$$

In addition, let g_x denote the posterior density of θ given that $X = x$. That is,

$$g_x(\theta) \equiv g(\theta|x) = \frac{f(x|\theta)g(\theta)}{f_g(x)}.$$

We shall need the following lemmas.

Lemma 5.1

$$g \underset{\text{l.r.}}{\geq} h \Rightarrow g_x \underset{\text{l.r.}}{\geq} h_x, \qquad \text{for all } x.$$

Proof: We must show that

$$\frac{g_x(\theta)}{h_x(\theta)} = \frac{g(\theta)}{h(\theta)} \frac{f_h(x)}{f_g(x)} \quad \text{is increasing in } \theta,$$

but this follows by assumption because $g \geq_{\text{l.r.}} h$. $\square$

Lemma 5.2

$$g_x \underset{\text{l.r.}}{\geq} g_y \qquad \text{when } x \geq y.$$

Proof: We must show that, when $x \geq y$,

$$\frac{g_x(\theta)}{g_y(\theta)} = \frac{f(x \mid \theta)}{f(y \mid \theta)} \frac{f_g(y)}{f_g(x)}$$

increases in θ; or equivalently, that

$$\frac{f(x \mid \theta_1)}{f(y \mid \theta_1)} \geq \frac{f(x \mid \theta_2)}{f(y \mid \theta_2)}, \qquad \text{for } x \geq y \text{ and } \theta_1 \geq \theta_2,$$

this is just condition (5.1). $\square$

Our final lemma states that if $g(\theta)/h(\theta)$ increases in θ, then f_g is a stochastically larger density than f_h.

Lemma 5.3 If $g \geq_{\text{l.r.}} h$, then

$$\int_x^\infty f_g(y)\, dy \geq \int_x^\infty f_h(y)\, dy, \qquad \text{for all } x.$$

Proof: We must show that $X_g \geq_{\text{st}} X_h$ when X_g and X_h are random variables having respective densities f_g and f_h. We shall use a coupling argument to generate such distributed random variables that $X_g \geq X_h$, which will prove the preceding lemma. Now because $g \geq_{\text{l.r.}} h$, it follows that $g \geq_{\text{st}} h$, so by coupling (see Section 2 of the Appendix), we can generate random variables θ_g and θ_h having respective densities g and h such that $\theta_g \geq \theta_h$. Now, because $f(\cdot \mid \theta_g) \geq_{\text{l.r.}} f(\cdot \mid \theta_h)$ implies that $f(\cdot \mid \theta_g) \geq_{\text{st}} f(\cdot \mid \theta_h)$, when $\theta_g \geq \theta_h$ we can generate random variables X_g and X_h from these respective distributions such that $X_g \geq X_h$. Because X_g and X_h have respective (unconditional) densities f_g and f_h, the result follows. $\square$

We are now ready to consider the classical optimization problem. Let us start by taking the state to be the posterior probability density g of θ, given all observations to that time. Letting $V(g)$ denote the optimal discounted return function starting in state g, we have the following optimality equation:

$$V(g) = \max\left[M, \mu(f_g) + \alpha \int V(g_x) f_g(x)\, dx \right]$$

where $\mu(f_g)$ is the mean of f_g, that is,

$$\mu(f_g) = \int x f_g(x)\, dx.$$

We shall now prove our main structural result, namely, that if g is greater than h in the sense of likelihood ratio, then $V(g) \geq V(h)$.

Proposition 5.4   If $g \geq_{\text{l.r.}} h$, then $V(g) \geq V(h)$.

Proof: Let V_k denote the maximal expected discounted return in a k-stage problem. Then

$$V_1(g) = \max\left[M, \mu(f_g) \right]$$

$$V_k(g) = \max\left[M, \mu(f_g) + \alpha \int V_{k-1}(g_x) f_g(x)\, dx \right], \qquad k > 1.$$

We now argue by induction that $g \geq_{\text{l.r.}} h$ implies that $V_k(g) \geq V_k(h)$ for all k. Because $g \geq_{\text{l.r.}} h$ implies from Lemma 5.3 that $f_g \geq_{\text{st}} f_h$ and thus $\mu(f_g) \geq \mu(f_h)$, the result is true for $k = 1$. So assume it for $k - 1$. To prove that $V_k(g) \geq V_k(h)$ it clearly suffices to show that

$$\int V_{k-1}(g_x) f_g(x)\, dx \geq \int V_{k-1}(h_x) f_h(x)\, dx.$$

Now, $V_{k-1}(g_x) \geq V_{k-1}(h_x)$ by Lemma 5.1 and the induction hypothesis, so it suffices to prove that

$$\int V_{k-1}(h_x) f_g(x)\, dx \geq \int V_{k-1}(h_x) f_h(x)\, dx.$$

However, it follows by Lemma 5.2 and the induction hypothesis that $V_{k-1}(h_x)$ is an increasing function of x, and because $f_g \geq_{\text{st}} f_h$ (by Lemma 5.3), the result follows.   □

So far we have suppressed, in our notation, the dependence on M. Let us now write $V(g:M)$ for $V(g)$. If we let $M(g)$ denote the indifference value for state g, that is,

$$M(g) = \min[M : V(g:M) = M],$$

then it follows from Proposition 5.4 that if $g \geq_{1.r.} h$ and $V(g:M) = M$, then $V(h:M) = M$. Hence, we have the following corollary.

Corollary 5.5 If $g \geq_{1.r.} h$, then $M(g) \geq M(h)$.

Consider now a multiarm bandit problem consisting of n independent projects. Suppose that the returns from project i have density $f(x|\theta_i)$, which satisfies (5.1), where $\theta_1, \ldots, \theta_n$ are regarded as independent random variables having respective prior densities $g_1, \ldots, g_n$. It follows from Theorem 3.3 that the optimal policy when the termination reward is M is to sample from that project having the maximal value of $M(g_i)$ if this maximal value exceeds M, and to retire otherwise. Hence, if one of the g_i exceeds the others in the sense of likelihood ratio, then it follows from Corollary 5.5 that either this project should be sampled or the process terminated. That is, we have the next corollary.

Corollary 5.6 If g_i is the posterior density of θ_i, $i = 1, \ldots, n$, and $g_i \geq_{1.r.} g_j$, $j = 1, 2, \ldots, n$, then it is optimal to sample from project i if $M(g_i) \geq M$, and to retire if $M(g_i) < M$.

Remark: If retirement is not an allowable option in the multi-project case, just let M be small enough so that it is never optimal to retire. For instance, if all returns are positive take $M = 0$.

Remark: From a computational point of view this choice of a state space is not very tractable. It is better to take the initial prior distribution g^0 as a parameter of the problem and then let the state be the vector of observed values. In fact, it often occurs (if so-called sufficient statistics exist) that the state can be taken to be some function of this vector. For instance, in the case where

$$f(1|\theta) = \theta \quad \text{and} \quad f(-1)|\theta) = 1 - \theta,$$

the posterior distribution of θ, given the observed values $x_1, \ldots, x_k$, can be shown (see Problem 7) to depend on $x_1, \ldots, x_k$ only through the number of them that are equal, respectively, to 1 and -1. Hence, the state space can be taken as the set of pairs (n,m), $n \geq 0$, $m \geq 0$, with the interpretation that the state is (n,m) if there have been n plus 1 observations out of a total of $n + m$ observations. As the posterior distribution of θ in state (n,m) increases in likelihood ratio in n and decreases in likelihood ratio in m (see Problem 7), it follows from Proposition 5.4 that

$$V(n, m + 1) \leq V(n, m) \leq V(n + 1, m),$$

and from Corollary 5.5 that

$$M(n, m + 1) \leq M(n, m) \leq M(n + 1, m).$$

The second part of this inequality implies that if the optimal policy wins with a given arm, then it will replay that arm on the next stage. That is, it will "stick with a winner."

Problems

1. If for functions $b(a)$ and $c(a)$, f is given by $f(x) = \max_a[b(a) + c(a)x]$, prove that $f(x)$ is convex in x.

2. Prove that, for any nonnegative integer-valued random variable Z,

$$\sum_{m=1}^{k} P(Z \geq m)\alpha^{m-1} = \sum_{m=1}^{k} P(Z = m)(1 + \alpha + \cdots + \alpha^{m-1})$$

$$+ P(Z > k)(1 + \alpha + \cdots + \alpha^{k-1}).$$

3. If X_j has discrete increasing failure rate, show that

$$\sum_{m=1}^{k} P(X_j = i + m)\alpha^{m-1} / \sum_{m=1}^{k} P(X_j \geq i + m)\alpha^{m-1} \quad \text{is increasing in } k.$$

4. If X_j has discrete increasing failure rate, show that $M^j(i)$ of Example 3.1 increases in i.

5. When X_j has discrete decreasing failure rate, show that the expression in Problem 3 is decreasing in k.

6. State and prove the equivalent of Theorem 3.3 when projects have restricted states.

7. Let $g(\theta)$ denote a prior density on θ, $0 < \theta < 1$, and suppose that $f(1|\theta) = \theta = 1 - f(-1|\theta)$.

(a) Show that the posterior density of θ, given $X_1, \ldots, X_{n+m}$, depends on $X_1, \ldots, X_{n+m}$ only through (n, m), where n is the number of the X_i that equal 1.

(b) Show that the posterior distribution of θ, given the outcome (n, m), increases in likelihood ratio in n and decreases in likelihood ratio in m.

Notes and References

The important result that multiproject bandit problems can be solved by considering the single-project problems along with a retirement return is due to Gittins (see, for instance [2]), though the approach we have followed in proving this result is from Whittle [3]. An extension to the case in which new projects arise according to a Poisson process is given in Whittle [4]. The classical single-arm bandit process is due to Bellman [1]. The small generalization of the classical model, presented in Section 4, is new.

1. Bellman, R., "A problem in the sequential design of experiments," *Sankhyā* **16,** 221–229, 1956.

2. Gittins, J. C., "Bandit processes and dynamic allocation indices," *J. Roy. Statist. Soc. Ser. B* **14,** 148–177, 1979.

3. Whittle, P., "Multi-armed bandits and the Gittins index," *J. Roy. Statist. Soc. Ser. B* **42,** 143–149, 1980.

4. Whittle, P., "Arm acquiring bandits," *Ann. Probab.* **9,** 284–292, 1981.

Appendix

Stochastic Order Relations

1. Stochastically Larger

We say that the random variable X is stochatically larger than the random variable Y, written $X \geq_{st} Y$, if

$$P(X > a) \geq P(Y > a) \qquad \text{for all } a. \tag{1.1}$$

Lemma 1.1 If $X \geq_{st} Y$, then $E(X) \geq E(Y)$.

Proof : Assume first that X and Y are nonnegative random variables. Then

$$E(X) = \int_0^\infty P(X > a)\,da \geq \int_0^\infty P(Y > a)\,da = E(Y).$$

In general, we can write any random variable Z as the difference of two nonnegative random variables as

$$Z = Z^+ - Z^-,$$

where

$$Z^+ = \begin{cases} Z & \text{if } Z \geq 0, \\ 0 & \text{if } Z < 0, \end{cases} \qquad Z^- = \begin{cases} 0 & \text{if } Z \geq 0, \\ -Z & \text{if } Z < 0. \end{cases}$$

We leave it as an exercise to show that

$$X \underset{st}{\geq} Y \Rightarrow X^+ \underset{st}{\geq} Y^+, \qquad X^- \underset{st}{\leq} Y^-.$$

153

Hence,

$$E(X) = E(X^+) - E(X^-) \geq E(Y^+) - E(Y^-) = E(Y). \quad \square$$

The next proposition gives an alternative definition of stochastically larger.

Proposition 1.2 $X \geq_{st} Y \Leftrightarrow E(f(X)) \geq E(f(Y))$ for all increasing functions f.

Proof : Suppose first that $X \geq_{st} Y$, and let f be an increasing function. We show that $f(X) \geq_{st} f(Y)$ as follows: Letting $f^{-1}(a) = \inf\{x : f(x) > a\}$, then

$$P[f(X) > a] = P[X > f^{-1}(a)] \geq P[Y > f^{-1}(a)] = P[f(Y) > a].$$

Hence, $f(X) \geq_{st} f(Y)$, so, from Lemma 1.1, $E(f(X)) \geq E(f(Y))$.

Now suppose that $E(f(X)) \geq E(f(Y))$ for all increasing functions f. For any a, let f_a denote the increasing function

$$f_a(x) = \begin{cases} 1 & \text{if } x > a, \\ 0 & \text{if } x \leq a, \end{cases}$$

then

$$E(f_a(X)) = P(X > a) \quad \text{and} \quad E(f_a(Y)) = P(Y > a),$$

and we see that $X \geq_{st} Y$. $\square$

2. Coupling

If $X \geq_{st} Y$, then there exists random variables X^* and Y^* having the same distributions as X and Y and such that X^* is, with probability 1, at least as large as Y^*. Before proving this we need the following lemma.

Lemma 2.1 Let F and G be continuous distribution functions. If X has distribution F, then the random variable $G^{-1}(F(X))$ has distribution G.

Proof :

$$P[G^{-1}(F(X)) \leq a] = P[F(X) \leq G(a)]$$
$$= P[X \leq F^{-1}(G(a))]$$
$$= F(F^{-1}(G(a))$$
$$= G(a) \quad \square$$

Proposition 2.2 If F and G are distributions such that $\bar{F}(a) \geq \bar{G}(a)$, then there exist random variables X and Y having distributions F and G, respectively, such that

$$P(X \geq Y) = 1.$$

Proof: We shall present a proof when F and G are continuous distribution functions. Let X have distribution F and define Y by $Y = G^{-1}(F(X))$. Then, by Lemma 2.1, Y has distribution G. But because $F \leq G$, it follows that $F^{-1} \geq G^{-1}$, so

$$Y = G^{-1}(F(X)) \leq F^{-1}(F(X)) = X,$$

which proves the result. $\square$

Often, the easiest way to prove that $\bar{F} \geq \bar{G}$ is by letting X be a random variable having distribution F and then defining a random variable Y in terms of X such that (i) Y has distribution G and (ii) $Y \leq X$. We illustrate this method, known as *coupling*, by some examples.

EXAMPLE 2.1 *Stochastic Ordering of Vectors* Let $X_1, \ldots, X_n$ be independent and $Y_1, \ldots, Y_n$ be independent. If $X_i \geq_{st} Y_i$, then for any increasing f,

$$f(X_1, \ldots, X_n) \underset{st}{\geq} f(Y_1, \ldots, Y_n).$$

Proof: Let $X_1, \ldots, X_n$ be independent, and use Proposition 2.2 to generate independent $Y_1^*, \ldots, Y_n^*$, where Y_i^* has the distribution of Y_i and $Y_i^* \leq X_i$. Then $f(X_1, \ldots, X_n) \geq f(Y_1^*, \ldots, Y_n^*)$ because f is increasing. Hence, for any a,

$$f(Y_1^*, \ldots, Y_n^*) > a \Rightarrow f(X_1, \ldots, X_n) > a,$$

so

$$P[f(Y_1^*, \ldots, Y_n^*) > a] \leq P[f(X_1, \ldots, X_n) > a].$$

Because the left side of the preceding expression is equal to

$$P[f(Y_1, \ldots, Y_n) > a],$$

the result follows.   □

EXAMPLE 2.2 *Stochastic Ordering of Poisson Random Variables.* We shall show that a Poisson random variable is stochastically increasing in its mean. For $\lambda_1 < \lambda_2$, let X_1 and X_2 be independent Poisson random variables with respective means λ_1 and $\lambda_2 - \lambda_1$. Now, $X_1 + X_2 \geq X_1$, and $X_1 + X_2$ is Poisson with mean λ_2, which proves the result.

3. Hazard-Rate Ordering

Let X be a nonnegative random variable with distribution F and density f. The failure (or hazard) rate function of X is defined by

$$\lambda(t) = \frac{f(t)}{\bar{F}(t)}, \qquad \bar{F}(t) = 1 - F(t).$$

The random variable X has a larger hazard (or failure) rate function than does Y if

$$\lambda_X(t) \geq \lambda_Y(t), \qquad \text{for all } t \geq 0, \tag{3.1}$$

where $\lambda_x(t)$ and $\lambda_Y(t)$ are the hazard rate functions of X and Y. Equation (3.1) states that, at the same age, the unit whose life is X is more likely to instantaneously perish than the one whose life is Y. In fact, because

$$P(X > t + s \mid X > t) = \exp\left[-\int_t^{t+s} \lambda(y)\, dy\right],$$

it follows that (3.1) is equivalent to

$$P(X > t + s \mid X > t) \leq P(Y > t + s \mid Y > t)$$

or, equivalently,

$$X_t \underset{st}{\leq} Y_t, \qquad \text{for all } t \geq 0,$$

where X_t and Y_t are, respectively, the remaining lives of a t-unit old item having the same distributions as X and Y.

4. Likelihood-Ratio Ordering

Let X and Y denote continuous nonnegative random variables having densities f and g, respectively. We say that X is larger than Y in the sense of likelihood ratio and write

$$X \underset{\text{LR}}{\geq} Y$$

if

$$\frac{f(x)}{g(x)} \leq \frac{f(y)}{g(y)}, \qquad \text{for} \quad \text{all } x \leq y.$$

Hence, $X \geq_{\text{LR}} Y$ if the ratio of their respective densities, $f(x)/g(x)$, is nondecreasing in x. We start by noting that this is a stronger ordering than failure-rate ordering (which is itself stronger than stochastic ordering).

Proposition 4.1   Let X and Y be nonnegative random variables having densities f and g and hazard-rate functions λ_X and λ_Y. If

$$X \underset{\text{LR}}{\geq} Y$$

then

$$\lambda_X(t) \leq \lambda_Y(t), \qquad \text{for} \quad \text{all } t \geq 0.$$

Proof: Because $X \geq_{\text{LR}} Y$, it follows that, for $x \geq t$,

$$f(x) \geq g(x) f(t) / g(t).$$

Hence,

$$\lambda_X(t) = \frac{f(t)}{\displaystyle\int_t^\infty f(x)\, dx}$$

$$\leq \frac{f(t)}{\displaystyle\int_t^\infty g(x) f(t)/g(t)\, dx}$$

$$= \frac{g(t)}{\displaystyle\int_t^\infty g(x)\, dx}$$

$$= \lambda_Y(t). \qquad \square$$

EXAMPLE 4.1a   If X is exponential with rate λ and Y is exponential with rate μ, then

$$\frac{f(x)}{g(x)} = \frac{\lambda}{\mu} e^{(\mu - \lambda)x},$$

so $X \geq_{LR} Y$ when $\lambda \leq \mu$.

EXAMPLE 4.1b   *A Statistical Inference Problem.*   A central problem in statistics is that of making inferences about the unknown distribution of a given random variable. In the simplest case, we suppose that X is a continuous random variable having a density function that is known to be either f or g. Based on the observed value of X, we must decide on either f or g.

A decision rule for this problem is a function $\phi(x)$, which takes on either value 0 or value 1 with the interpretation that if X is observed to equal x, then we decide on f if $\phi(x) = 0$ and on g if $\phi(x) = 1$. To help us decide on a good decision rule, let us first note that

$$\int_{x:\phi(x)=1} f(x)\, dx = \int f(x)\phi(x)\, dx$$

represents the probability of rejecting f when it is in fact the true density. The classical approach to obtaining a decision rule is to fix a constant α, $0 \leq \alpha \leq 1$, and then restrict attention to decision rules ϕ such that

$$\int f(x)\phi(x)\, dx \leq \alpha. \tag{4.1}$$

From among such rules it then attempts to choose the one that maximizes the probability of rejecting f when it is false. That is, it maximizes

$$\int_{x:\phi(x)=1} g(x)\, dx = \int g(x)\phi(x)\, dx.$$

The optimal-decision rule, according to this criterion, is given by the following proposition, known as the Neyman–Pearson lemma.

Neyman–Pearson Lemma   Among all decision rules ϕ satisfying

(4.1), the one that maximizes $\int g(x)\phi(x)\,dx$ is ϕ^*, given by

$$\phi^*(x) = \begin{cases} 0 & \text{if } \dfrac{f(x)}{g(x)} \geq c, \\[2mm] 1 & \text{if } \dfrac{f(x)}{g(x)} < c, \end{cases}$$

where c is chosen so that

$$\int f(x)\phi^*(x)\,dx = \alpha.$$

Proof: Let ϕ satisfy (4.1). Now, for any x,

$$[\phi^*(x) - \phi(x)][cg(x) - f(x)] \geq 0.$$

This inequality follows because if $\phi^*(x) = 1$, then both terms in the product are nonnegative, and if $\phi^*(x) = 0$, then both are nonpositive. Hence,

$$\int [\phi^*(x) - \phi(x)][cg(x) - f(x)]\,dx \geq 0,$$

so

$$c\left[\int \phi^*(x)g(x)\,dx - \int \phi(x)g(x)\,dx\right] \geq \int \phi^*(x)f(x)\,dx - \int \phi(x)f(x)\,dx$$

$$\geq 0,$$

which proves the result. $\square$

If we suppose now that f and g have a monotone likelihood-ratio order, that is, $f(x)/g(x)$ is nondecreasing in x, then the optimal decision rule can be written as

$$\phi^*(x) = \begin{cases} 0 & \text{if } x \geq k, \\ 1 & \text{if } x < k, \end{cases}$$

where k is such that

$$\int_{-\infty}^{k} f(x)\,dx = \alpha.$$

That is, the optimal decision rule is to decide on f when the observed value is greater than some critical number and to decide on g otherwise.

Problems

1. If $X \geq_{st} Y$, prove that $X^+ \geq_{st} Y^+$ and $Y^- \geq_{st} X^-$.

2. Suppose $X_i \geq_{st} Y_i$, $i = 1, 2$. Show by counterexample that it is not necessarily true that $X_1 + X_2 \geq_{st} Y_1 + Y_2$.

3. (a) If $X \geq_{st} Y$, show that $P(X \geq Y) \geq \frac{1}{2}$. *Assume independence.*
   (b) If $P(X \geq Y) \geq \frac{1}{2}$, $P(Y \geq Z) \geq \frac{1}{2}$, and X, Y, Z are independent, does this imply that $P(X \geq Z) \geq \frac{1}{2}$? Prove this or give a counterexample.

4. One of n elements will be requested. It will be i with probability P_i, $i = 1, \ldots, n$. If the elements are to be arranged in an ordered list, find the arrangement that stochastically minimizes the position of the element requested.

5. Show that a binomial n,p distribution $B_{n,p}$ increases stochastically both as n increases and as p increases. That is, $\bar{B}_{n,p}$ increases in n and in p.

6. Prove that a normal distribution with mean μ and variance σ^2 increases stochastically as μ increases; what about as σ^2 increases?

7. Consider a Markov chain with transition probability matrix P_{ij}, and suppose that $\sum_{j=k}^{\infty} P_{ij}$ increases in i for all k.
   (a) Show that, for all increasing functions f, $\sum_j P_{ij} f(j)$ increases in i.
   (b) Show that $\sum_{j=k}^{\infty} P_{ij}^n$ increases in i for all k, where P_{ij}^n are the n-step transition probabilities, $n \geq 2$.

8. Let X_1 and X_2 have respective hazard-rate functions $\lambda_1(t)$ and $\lambda_2(t)$. Show that $\lambda_1(t) \geq \lambda_2(t)$ for all t is and only if

$$P(X_1 > t)/(P(X_1 > s) \leq P\{X_2 > t\}/P(X_2 > s),$$

for all $s < t$.

9. Let F and G have hazard-rate functions λ_F and λ_G. Show that $\lambda_F(t) \geq \lambda_G(t)$ for all t if and only if there exist independent continuous random variables Y and Z such that Y has distribution G and $\min(Y, Z)$ has distribution F.

10. A family of random variables $\{X_\theta, \theta \in [a,b]\}$ is said to be a monotone likelihood-ratio family if

$$X_{\theta_1} \underset{LR}{\leq} X_{\theta_2} \qquad \text{when} \qquad \theta_1 \leq \theta_2.$$

Show that the following families have monotone likelihood ratio.

(a) X_θ is binomial with parameters n and θ, n fixed.

(b) X_θ is Poisson with mean θ.

(c) X_θ is uniform $(0, \theta)$.

(d) X_θ is gamma with parameters $(n, 1/\theta)$, n fixed.

(e) X_θ is gamma with parameters (θ, λ), λ fixed.

11. Consider the statistical inference problem in which a random variable X is known to have either density f or g. The Bayesian approach is to postulate a prior probability p that f is the true density. The hypothesis that f were the true density would then be accepted if the posterior probability given the value of X is greater than some critical number. Show that if $f(x)/g(x)$ is non-decreasing in x, this is equivalent to accepting f whenever the observed value of X is greater than some critical value.

Reference

1. Ross, S. M., *Stochastic Processes*, Wiley, New York, 1982.

Index

Probability and Mathematical Statistics

A Series of Monographs and Textbooks

Editors   **Z. W. Birnbaum**   **E. Lukacs**
University of Washington *Bowling Green State University*
Seattle, Washington   *Bowling Green, Ohio*

Thomas Ferguson. Mathematical Statistics: A Decision Theoretic Approach. 1967

Howard Tucker. A Graduate Course in Probability. 1967

K. R. Parthasarathy. Probability Measures on Metric Spaces. 1967

P. Révész. The Laws of Large Numbers. 1968

H. P. McKean, Jr. Stochastic Integrals. 1969

B. V. Gnedenko, Yu. K. Belyayev, and A. D. Solovyev. Mathematical Methods of Reliability Theory. 1969

Demetrios A. Kappos. Probability Algebras and Stochastic Spaces. 1969

Ivan N. Pesin. Classical and Modern Integration Theories. 1970

S. Vajda. Probabilistic Programming. 1972

Sheldon M. Ross. Introduction to Probability Models. 1972

Robert B. Ash. Real Analysis and Probability. 1972

V. V. Fedorov. Theory of Optimal Experiments. 1972

K. V. Mardia. Statistics of Directional Data. 1972

H. Dym and H. P. McKean. Fourier Series and Integrals. 1972

Tatsuo Kawata. Fourier Analysis in Probability Theory. 1972

Fritz Oberhettinger. Fourier Transforms of Distributions and Their Inverses: A Collection of Tables. 1973

Paul Erdös and Joel Spencer. Probabilistic Methods in Combinatorics. 1973

K. Sarkadi and I. Vincze. Mathematical Methods of Statistical Quality Control. 1973

Michael R. Anderberg. Cluster Analysis for Applications. 1973

W. Hengartner and R. Theodorescu. Concentration Functions. 1973

Kai Lai Chung. A Course in Probability Theory, Second Edition. 1974

L. H. Koopmans. The Spectral Analysis of Time Series. 1974

L. E. Maistrov. Probability Theory: A Historical Sketch. 1974

William F. Stout. Almost Sure Convergence. 1974

E. J. McShane. Stochastic Calculus and Stochastic Models. 1974

Robert B. Ash and Melvin F. Gardner. Topics in Stochastic Processes. 1975

Avner Friedman. Stochastic Differential Equations and Applications, Volume 1, 1975; Volume 2. 1975

Roger Cuppens. Decomposition of Multivariate Probabilities. 1975

Eugene Lukacs. Stochastic Convergence, Second Edition. 1975

H. Dym and H. P. McKean. Gaussian Processes, Function Theory, and the Inverse Spectral Problem. 1976

N. C. Giri. Multivariate Statistical Inference. 1977

Lloyd Fisher and John McDonald. Fixed Effects Analysis of Variance. 1978

Sidney C. Port and Charles J. Stone. Brownian Motion and Classical Potential Theory. 1978

Konrad Jacobs. Measure and Integral. 1978

K. V. Mardia, J. T. Kent, and J. M. Biddy. Multivariate Analysis. 1979

Sri Gopal Mohanty. Lattice Path Counting and Applications. 1979

Y. L. Tong. Probability Inequalities in Multivariate Distributions. 1980

Michel Metivier and J. Pellaumail. Stochastic Integration. 1980

M. B. Priestly. Spectral Analysis and Time Series. 1980

Ishwar V. Basawa and B. L. S. Prakasa Rao. Statistical Inference for Stochastic Processes. 1980

M. Csörgö and P. Révész. Strong Approximations in Probability and Statistics. 1980

Sheldon Ross. Introduction to Probability Models, Second Edition. 1980

P. Hall and C. C. Heyde. Martingale Limit Theory and Its Application. 1980

Imre Csiszár and János Körner. Information Theory: Coding Theorems for Discrete Memoryless Systems. 1981

A. Hald. Statistical Theory of Sampling Inspection by Attributes. 1981

H. Bauer. Probability Theory and Elements of Measure Theory. 1981

M. M. Rao. Foundations of Stochastic Analysis. 1981

Jean-Rene Barra. Mathematical Basis of Statistics. Translation and Edited by L. Herbach. 1981

Harald Bergström. Weak Convergence of Measures. 1982

Sheldon Ross. Introduction to Stochastic Dynamic Programming. 1983

in preparation:

B. L. S. Prakasa Rao. Nonparametric Functional Estimation.